VOLUME FIFTY FOUR

ADVANCES IN INSECT PHYSIOLOGY

Butterfly Wing Patterns and Mimicry

VOLUME FIFTY FOUR

ADVANCES IN INSECT PHYSIOLOGY

Butterfly Wing Patterns and Mimicry

Edited by

RICHARD H. ffRENCH-CONSTANT

Center for Ecology and Conservation,
University of Exeter, Penryn, United Kingdom

Series Editor

RUSSELL JURENKA

Department of Entomology,
Iowa State University, Iowa, United States

ELSEVIER

ACADEMIC PRESS

An imprint of Elsevier

Academic Press is an imprint of Elsevier
125 London Wall, London, EC2Y 5AS, United Kingdom
The Boulevard, Langford Lane, Kidlington, Oxford OX5 1GB, United Kingdom
525 B Street, Suite 1800, San Diego, CA 92101-4495, United States
50 Hampshire Street, 5th Floor, Cambridge, MA 02139, United States

First edition 2018

Notices
Knowledge and best practice in this field are constantly changing. As new research and experience broaden our understanding, changes in research methods, professional practices, or medical treatment may become necessary.

Practitioners and researchers must always rely on their own experience and knowledge in evaluating and using any information, methods, compounds, or experiments described herein. In using such information or methods they should be mindful of their own safety and the safety of others, including parties for whom they have a professional responsibility

To the fullest extent of the law, neither the Publisher nor the authors, contributors, or editors, assume any liability for any injury and/or damage to persons or property as a matter of products liability, negligence or otherwise, or from any use or operation of any methods, products, instructions, or ideas contained in the material herein.

ISBN: 978-0-12-813770-3
ISSN: 0065-2806

Publisher: Zoe Kruze
Acquisition Editor: Sam Mahfoudh
Editorial Project Manager: Alina Cleju
Production Project Manager: Abdulla Sait
Cover Designer: Victoria Pearson
Cover Image Courtesy: Richard H. ffrench-Constant

Typeset by SPi Global, India

CONTENTS

CONTRIBUTORS

Saad Arif
Oxford Brookes University, Oxford, United Kingdom

Nora Braak
Oxford Brookes University, Oxford, United Kingdom

Casper J. Breuker
Oxford Brookes University, Oxford, United Kingdom

Leandro Duarte
Universidade Federal do Rio Grande do Sul (UFRGS), Porto Alegre, Rio Grande do Sul, Brazil

Melanie Gibbs
NERC Centre for Ecology & Hydrology, Wallingford, United Kingdom

Cristiano A. Iserhard
Universidade Federal de Pelotas (UFPEL), Pelotas, Rio Grande do Sul, Brazil

Lucas A. Kaminski
Universidade Federal do Rio Grande do Sul (UFRGS), Porto Alegre, Rio Grande do Sul, Brazil

Luca Livraghi
Oxford Brookes University, Oxford, United Kingdom

Arnaud Martin
The George Washington University, Washington, DC, United States

Sébastien R. Mouchet
School of Physics, University of Exeter, Exeter, United Kingdom

Dirleane O. Rossato
Universidade Federal do Rio Grande do Sul (UFRGS), Porto Alegre, Rio Grande do Sul, Brazil

Pete Vukusic
School of Physics, University of Exeter, Exeter, United Kingdom

INTRODUCTORY CHAPTER: BUTTERFLY WING PATTERNS: NOT JUST PAINTING BY NUMBERS

Richard H. ffrench-Constant
Center for Ecology and Conservation, University of Exeter, Penryn, United Kingdom

Abstract

Despite our growing understanding of the genes underlying butterfly wing pattern formation and pigmentation, there is a growing gap in our understanding of how these two processes are linked in developmental time. Here I review the concept of the developmental heterochrony that drives the differential rates of scale development across the butterfly wing, which in turn are responsible for the final differences in their pigmentation. I discuss how this 'developmental gap' might be best filled and what gene products might occupy it. Despite early implications that ecdysone receptor expression must be involved in the differential read-out of the declining ecdysteroid titre in the pupa close to eclosion, no other genes linking these two processes have been identified. As scales in the wing initiate development, their progress must be driven by genes involved in the cell cycle which must in turn cause the differential scale development that we see across the developing wing. Key advances in our ability to genetically transform butterflies using CRISPR–CAS are highlighted and a potential road map suggested for the possible use of this and other techniques in filling the growing 'gap' between patterning and pigmentation.

1. PAINTING BY NUMBERS

Butterfly wing patterning is strictly a developmental problem and can be roughly divided into three key developmental phases: 1. *Pattern formation*, in which signalling molecules (often transcription factors) lay down the basic pattern. 2. *Differential scale cell maturation* (or so-called developmental 'heterochrony' (ffrench-Constant, 2012; Koch et al., 2000b)) in which scales fated to be different colours differentiate on different developmental trajectories. 3. *Pigmentation*, in which the different scale cells adopt their different colours via becoming either coloured (depending on which pigment precursors are available during their development) or melanised black/brown. If we use the analogy of a child 'colouring by numbers' (filling in predefined patterns with each colour specified by different numbers in a printed book), then pattern formation is akin to the printed book (with the numbers representing different colours) and pigmentation is likened to the child dipping his or her brush into pots of pigment in an order specified by those numbers. However, the central difference is that there is no differential development of any scale-like structures on the page determining the uptake and/or availability of pigments through time—the child simply dips his or her brush into a specific pot in a predetermined order. In other words the page is simply blank—it does not have scales that drive the final colour—the colours need to be applied by the child. Using this simple conceptual framework, we can now critically examine the literature to see what we know about each of these overlapping developmental steps and we can then identify the key 'research gaps' going forward. More importantly, we can then discuss how these might be filled.

2. MIND THE DEVELOPMENTAL GAP

Following the seminal work of Carroll and co-workers in defining the array of transcription factor expression that underlies eyespot determination in different butterflies (Carroll, 1998; Keys et al., 1999; Nijhout, 1980), many different transcription factors (Reed et al., 2011) and also signalling molecules, such as WntA (Mazo-Vargas et al., 2017), have been described as essential elements in defining the pattern. Returning to our analogy, this is like the printing and numbering the blank spaces on the page that the child will later have to colour-in. Correspondingly, following the excellent review of butterfly wing patterning by Frederick Nijhout in his classic book, we now also know much more about the chemical make-up of the different pigments butterflies and moths use and even some of the key enzymes that determine their deposition in the developing scale cells. The beginning (pattern formation) and end (pigmentation) of wing pattern development therefore seem relatively well understood. However, the central problem is the huge inferred 'gap' between the transcription factors that lay down the pattern and the pigmentation genes that derive the final scale cell colours. Despite the fact that we know that pigmentation is performed within the pupa and close to eclosion (hatching) and is set against a steadily falling titre of ecdysteroids, we *don't understand how the scale cells respond to this falling hormone titre and why scale cells develop at different rates* (developmental heterochrony) *across the wing.*

3. WHAT IS MISSING?

What we lack is a basic understanding of scale cell development and how this drives the final colour derived. Scales that develop 'early' tend to be coloured with pigments specific to the type of butterfly involved, whereas scales that develop later are melanised black or brown (Koch et al., 2000a). Key work by Koch et al. (2003) has already hinted that differential display of ecdysone receptors within the developing eyespot is central to the response of the developing scale to the falling ecdysteroid titre close to eclosion. Moreover, this differential expression of ecdysone receptors must somehow trigger the differential development that is the hallmark of scale cell heterochrony. But what are the other missing steps? How does microtubule development determine the shape and development of different scales (Dinwiddie et al., 2014)? What pigment transporters are available when and where to take up any necessary precursors from the surrounding haemolymph? What drives the final sclerotisation (cuticle hardening) of the scale cell and how is this related to its final colour (coloured or melanised black of brown?).

One key finding here is that the genes that seem to be under natural selection, in other words the genes that are found at colour pattern loci (mapped via genetic mapping), also appear to be involved in the *cell cycle*. Thus a *cortex*-like gene appears to control melanism in both *Heliconius* (the presence or absence of the yellow hindwing band encoded by the *Yb* locus (Nadeau et al., 2016)) and *Biston betularia* the Peppered moth (which is the classic example of industrial melanism (van't Hof et al., 2011)). The finding that *cortex*-like cell cycle controlling genes are central to controlling the fate of white/yellow scales vs their black counterparts is a key indicator as to how the different developmental trajectories of coloured (pigmented) and black (melanised) scale cells are set into motion. Alternatively we need to go back to working on *Drosophila* to further investigate the clear parallels between butterfly scales and fly hairs as revealed by genes such as *achaete-scute* (Galant et al., 1998).

4. HOW CAN NEW TOOLS HELP?

This volume brings together key current studies that begin to address this developmental gap. One key advance is the development of CRISPR–CAS-based gene technology, reviewed by Casper Breuker and co-workers in this volume, to perform target knockouts within butterflies and moths. This now actually allows us to target candidate genes (perhaps identified via genome-wide association or biochemical studies) and to knock them out or down in order to directly look at their function in the final adult wing pattern. While CRISPR–CAS in butterflies is still most efficient at generating tissue mosaics (patches of transformed and untransformed tissue on the wing), further progress and an understanding of exactly which tissues need to be injected should ensure stable germline transformants. CRISPR–CAS-mediated germline transformants are genetically stable and can be kept as reference butterfly stocks. This will be a major step forward and should then allow us to work systematically backwards from pigmentation and forwards from transcription factor expression to fill the knowledge gap between.

5. PATTERN IS NOT EVERYTHING

As we learn more and more about the 'static' patterns that are found across the wings of different butterflies and moths and the genes that underlie their formation, we are also made aware that insects are rarely static themselves! Therefore, if aposematic patterns are involved in mimicry, butterflies and moths may also want to mimic each other's wing *shapes* and/or *flight behaviour*. This aspect of mimicry is often largely overlooked and there are few studies on how wing shape is altered to effect better mimicry, with *Heliconius numata* being a notable exception (Jones et al., 2013). But of course it is not only the shape of the wings that is important but also the flight behaviour of the butterfly including its wing beat frequency. The role of wing shape across butterflies is comprehensively reviewed here, perhaps for the first time, and the study of the motion behind mimicry should be a major growth area in future research.

6. WHERE DO WE GO FROM HERE?

Although techniques such as CRISPR–CAS threaten to revolutionise our understanding of the role of genes in butterfly wing patterning, another key approach is to simply resequence butterflies of different colour morphs in order to look for genes associated with pattern via techniques such as GWAS (genome-wide association studies). These techniques have the advantage of being assumption free and have been key to out ability to be able to start to bridge the patterning to pigmentation gene gap. So even if we fail to logically fill in the network of genes between patterning and pigmentation, more and more candidates will arise from population genomic studies of new model butterflies and moths. These are exciting times in this field, and within another 10 years, we should have a much better idea of how developmental heterochrony of scale cells is set up and by which key genes it is driven.

ACKNOWLEDGEMENTS

I would like to thank Alina Cleju at Elsevier for managing this project. Work on butterflies in the ffrench-Constant laboratory is supported by the BBSRC, the Leverhulme Trust and the Royal Society of London (Merit Award to R.H.ff.-C).

REFERENCES

Carroll, S.B., 1998. Migrating to butterflies. Curr. Biol. 8 (15), R511–R512.

Dinwiddie, A., Null, R., Pizzano, M., et al., 2014. Dynamics of F-actin prefigure the structure of butterfly wing scales. Dev. Biol. 392 (2), 404–418. https://doi.org/10.1016/j.ydbio.2014.06.005.

ffrench-Constant, R.H., 2012. Butterfly wing colours are driven by the evolution of developmental heterochrony. Butterfly wing colours and patterning by numbers. Heredity (Edinb.) 108 (6), 592–593. https://doi.org/10.1038/hdy.2012.6.

Galant, R., Skeath, J.B., Paddock, S., Lewis, D.L., Carroll, S.B., 1998. Expression pattern of a butterfly achaete-scute homolog reveals the homology of butterfly wing scales and insect sensory bristles. Curr. Biol. 8 (14), 807–813.

Jones, R.T., Le Poul, Y., Whibley, A.C., Merot, C., ffrench-Constant, R.H., Joron, M., 2013. Wing shape variation associated with mimicry in butterflies. Evolution 67 (8), 2323–2334. https://doi.org/10.1111/evo.12114.

Keys, D.N., Lewis, D.L., Selegue, J.E., et al., 1999. Recruitment of a hedgehog regulatory circuit in butterfly eyespot evolution. Science 283 (5401), 532–534.

Koch, P.B., Behnecke, B., ffrench-Constant, R.H., 2000a. The molecular basis of melanism and mimicry in a swallowtail butterfly. Curr. Biol. 10 (10), 591–594.

Koch, P.B., Lorenz, U., Brakefield, P.M., ffrench-Constant, R.H., 2000b. Butterfly wing pattern mutants: developmental heterochrony and co-ordinately regulated phenotypes. Dev. Genes Evol. 210 (11), 536–544. https://doi.org/10.1007/s004270050344.

Koch, P.B., Merk, R., Reinhardt, R., Weber, P., 2003. Localization of ecdysone receptor protein during colour pattern formation in wings of the butterfly Precis coenia (Lepidoptera: Nymphalidae) and co-expression with Distal-less protein. Dev. Genes Evol. 212 (12), 571–584. https://doi.org/10.1007/s00427-002-0277-5.

Mazo-Vargas, A., Concha, C., Livraghi, L., et al., 2017. Macroevolutionary shifts of WntA function potentiate butterfly wing-pattern diversity. Proc. Natl. Acad. Sci. U.S.A. 114 (40), 10701–10706. https://doi.org/10.1073/pnas.1708149114.

Nadeau, N.J., Pardo-Diaz, C., Whibley, A., et al., 2016. The gene cortex controls mimicry and crypsis in butterflies and moths. Nature 534 (7605), 106–110. https://doi.org/10.1038/nature17961.

Nijhout, H.F., 1980. Pattern formation on lepidopteran wings: determination of an eyespot. Dev. Biol. 80 (2), 267–274.

Reed, R.D., Papa, R., Martin, A., et al., 2011. optix drives the repeated convergent evolution of butterfly wing pattern mimicry. Science 333 (6046), 1137–1141. https://doi.org/10.1126/science.1208227.

van't Hof, A.E., Edmonds, N., Dalikova, M., Marec, F., Saccheri, I.J., 2011. Industrial melanism in British peppered moths has a singular and recent mutational origin. Science 332 (6032), 958–960. https://doi.org/10.1126/science.1203043.

CHAPTER ONE

Structural Colours in Lepidopteran Scales

Sébastien R. Mouchet, Pete Vukusic
School of Physics, University of Exeter, Exeter, United Kingdom

Contents

Advances in Insect Physiology, Volume 54
ISSN 0065-2806
https://doi.org/10.1016/bs.aiip.2017.11.002

Abstract

Photonic structures incorporated in lepidopteran scales are responsible for a very broad range of optical effects: iridescence, narrow-band reflection, large solid-angle scattering, polarisation effects, additive colour mixing and more. They have been the most investigated natural photonic structures for a long time. Such studies provide both understanding of the optical mechanisms and the biological functions behind these effects as well as inspiration for the design and development of novel photonic materials through a bioinspiration approach. In this chapter, research regarding structural colours in lepidopteran wing scales is reviewed through the classification of the related photonic structures. Selected examples of these structures are used to illustrate how such optical effects are brought about.

1. INTRODUCTION

Colour and the mechanisms by which they are produced have always fascinated mankind in various ways. Early hominids used bright colour palettes in cave paintings; ancient and more recent civilisations incorporated colour into ceramics, textiles, art, jewellery and even food preparation. Underpinning this, the connecting thread for the curious mind and scientific bent has been a search for understanding dramatically colourful natural phenomena, such as the saturation of skies at sunset and sunrise, the colours of plant flowers and tree leaves across seasons, the whiteness of clouds and snow and the dazzling display of rainbows and auroras.

In nature, many living organisms, ranging from plants to mammals and birds, have very conspicuous appearances and exhibit striking optical effects (Berthier, 2003; Kinoshita, 2008; Vukusic and Sambles, 2003). These colour effects often facilitate inter- and intraspecific communication, e.g., conspecific visual recognition, courtship and aposematism. They can also play a role in camouflage or thermoregulation. As a result of these properties' strong significance in species' survival and courtship behaviours, they have been subject to potent selection pressures.

While pigmentation gives rise to the majority of colours in biological systems, it is clear too that a large variety of physical structures, at the scale of around a hundred to a few hundred nanometres or so, give rise to many strong and especially conspicuous optical effects and colourful appearances. In the case of some of the most eye-catching colourful insects, for example, such as Lepidoptera, pigmentation and colour-generating nanostructures are usually located in scales covering both surfaces of their wings (Fig. 1). These

Fig. 1 (A) The Purple Emperor *Apatura iris* butterfly displays a *brown colour* appearance with an iridescent shine ranging from *violet* to *near-UV*. (B, C) Two kinds of scales cover its wings: cover scales and ground scales as observed by scanning electron microscopy (SEM) (B) and transmission electron microscopy (TEM) (C). The ridge periodicities are ca. 1 and 1.5 µm, respectively. Only the cover scales (D) give rise to iridescent UV–violet reflection, in contrast to the ground scales (E). The period of the lamella stacking is equal to about 150 nm. The *brown colour* is due to melanin pigmentation.

scales develop in the chrysalis from epidermal cells based in the epithelial layer of the wings (Ghiradella, 1985, 1989, 1998). Two layers of scales (sometimes three) imbricated on the wings usually form a continuous set of pointillist colour sites that can give the macroscopic appearance of colour and pattern.

In this chapter, we review and classify the different photonic nanostructures occurring in lepidopteran wing scales and relate them to their optical responses.

1.1 Human Colour Vision

When describing scientifically a colour, it is useful to use a quantitative tool that is more elaborate than the wavelength of peak reflectance and the related full width at half maximum. It can be complex to infer the colour appearance of a surface from a cursory inspection of its reflection spectrum.

Colourimetry is the science and technology involved in the quantification of colour. Several specific systems, known as colour spaces, have been developed (Chamberlin and Chamberlin, 1980; Judd and Wyszecki, 1975) for different applications in arts such as photography or in industrial fields related to glass, metal, textile, food, cosmetics, plastics, paper or ink industries.

The colour space that is widely used in the characterisation of structural colours in living organisms was defined by the International Commission on Illumination (abbreviated CIE) in 1931. The CIE 1931 colour space is the result of the first international agreement on the mathematical processing of colour appearance. This colour space system is only valid for human vision. Other faunal visual systems have different spectral sensitivities, with some also sensitive to polarised light.

In this colour space, any colour can be quantified by chromaticity coordinates (x, y, z), calculated from a reflectance or transmittance spectrum $S(\lambda)$ using the relationships:

$$
\begin{aligned}
X &= k \int S(\lambda) L(\lambda) \overline{x}(\lambda) d\lambda \\
Y &= k \int S(\lambda) L(\lambda) \overline{y}(\lambda) d\lambda, \\
Z &= k \int S(\lambda) L(\lambda) \overline{z}(\lambda) d\lambda
\end{aligned}
\tag{1}
$$

where X, Y and Z are referred to as the tristimulus values; $k = 100 / \int L(\lambda)\overline{y}(\lambda) d\lambda$ is a normalisation constant defined so that an object exhibiting an optical response $S(\lambda) = 1$ for every wavelength λ gives rise to a Y value equal to 100; $L(\lambda)$ is the spectrum associated with the light source and $\overline{x}(\lambda)$, $\overline{y}(\lambda)$ and $\overline{z}(\lambda)$ are CIE colour-matching functions describing numerically the chromatic sensitivity of typical human vision (Chamberlin and Chamberlin, 1980; Judd and Wyszecki, 1975). The integral limits are commonly chosen to be 380 and 780 nm. Often, the incident light source spectrum is the D_{65} illuminant, which corresponds to the average daylight emitted by the Sun (assumed to be a 6500 K black body) and shed on Northern Europe.

The chromaticity coordinates are given by:

$$
\begin{aligned}
x &= \frac{X}{X+Y+Z} \\
y &= \frac{Y}{X+Y+Z}. \\
z &= \frac{Z}{X+Y+Z}
\end{aligned}
\tag{2}
$$

The two independent coordinates (x, y) define the chromaticity on the CIE diagram.

The CIE also defined an RGB colour space. The red (r), green (g) and blue (b) chromaticity coordinates can be calculated from $\bar{r}(\lambda)$, $\bar{g}(\lambda)$ and $\bar{b}(\lambda)$ colour-matching functions defined by the CIE or from the X, Y and Z tristimulus values by using a conversion matrix (Chamberlin and Chamberlin, 1980; Judd and Wyszecki, 1975). The obtained R, G and B tristimulus values then provide (r, g, b) coordinates after normalisation.

1.2 Origins of Colours in Nature

Colours in nature are classified into three categories, according to their origin: pigmentary colours, due to selective absorption of incident light by pigments; colours resulting from light emission through processes such as fluorescence, phosphorescence or bioluminescence; and structural colours that arise from the interaction of light with physical structures. Most colours in living organisms such as plants or insects belong to the first category. In the biological world, it is not uncommon to find organisms using a combination of two or even three of these colouration mechanisms in order to give rise to their specific appearance. Some examples will be presented in Section 4.

Pigments comprise naturally occurring chemicals that are coloured and are formed in mineral or biological (vegetal or animal) systems. Melanin, pterin, ommochrome and papiliochrome commonly colour in many insects (Berthier, 2003). Often, these molecules act as filters, absorbing short wavelengths by electron excitation so that unabsorbed light scattered at longer wavelengths creates the colour. In butterfly wings for example, melanin is often associated with black and brown colour appearances; papiliochrome R, with reddish-brown patches; papiliochrome II with yellow scales; 3-hydroxykynurenine (3-OHK), with orange and red colour appearances; xanthopterin and erythropterin with yellow, orange and red colour appearances (Berthier, 2003; Morehouse et al., 2007; Wilts et al., 2012c). Finally, pigments such as leucopterin absorb in the UV range, which does not affect the colour perceived by human eyes but which is very different to colour appearance perceived by butterflies' visual systems (Makino et al., 1952; Yagi, 1954).

Fluorescence emission occurs in many living organisms (Lagorio et al., 2015; Marshall and Johnsen, 2017): insects (Cockayne, 1924; Israelowitz et al., 2007; Welch et al., 2012), arachnids (Lawrence, 1954; Pavan and Vachon, 1954), amphibians (Taboada et al., 2017), marine animals

(Catala-Stucki, 1959; Phillips, 1927), birds (Arnold et al., 2002; McGraw et al., 2007), plants (Goodwin, 1953) or mammals (Tani et al., 2004). Fluorophores, such as biopterin, papiliochrome II or green fluorescent protein (GFP), embedded within biological tissues, emit visible light upon ultraviolet (UV) illumination: this gives rise to a colour appearance that, depending on the fluorophore, can be associated with any part of the visible spectrum.

Structural colours can arise due to the interaction of light with nanostructures: these are also referred to as photonic structures and often comprise elements that exhibit alternating high and low refractive indices (RI). These nanostructures are often relatively periodic and have dimensions on the order of visible light wavelengths, i.e., a few hundred nanometres. Biological structural colours have been associated with a range of living organisms, some dating to more than 500 million years (Kinoshita, 2008; Parker, 2000, 2005): insects (Berthier, 2003, 2010; Biró and Vigneron, 2011; Seago et al., 2009), arachnids (Ingram et al., 2009, 2011; Oxford and Gillespie, 1998), birds (Prum et al., 1998; Vigneron et al., 2006; Yin et al., 2006; Yoshioka and Kinoshita, 2002), fish (Denton, 1971; Mäthger et al., 2003), molluscs (Holt et al., 2014; Tan et al., 2004), mammals (Prum and Torres, 2004) or plants (Lee and Lowry, 1975; Vignolini et al., 2012). Examples of structures, detailed in Section 3, are found in the forms of thin films, multilayers, diffraction gratings, photonic crystals (PCs) and quasi- or randomly disordered structures. Many of these structures give rise to discernible iridescence, i.e., colour change with the incidence and observation angles. Another feature of structural colour is its resilience to ageing, in contrast to pigment colour that photobleaches with exposure to strong sunlight.

Natural nanostructures are often porous and comprise biopolymers such as chitin (in arthropods), keratin (in birds), collagen (in mammals) or cellulose (in plants), in addition to air, water or even physiological liquids. In insects, they are usually located in their cuticle, i.e., the outer covering layer of their bodies, or in scales covering their bodies and wings. Chitin is usually believed to be the main building material of these structures in insects. It is a polysaccharide, the chemical formula of which is $(C_8H_{13}O_5N)_n$, that forms rods within a protein matrix. It was chemically isolated for the first time by the French chemist Henri Braconnot in 1811 and was the first identified polysaccharide, described in detail 30 years before cellulose. Found in the integuments of many living organisms, its name comes from the Greek word *Χιτών* that means "covering". It is the second most abundant biopolymer on Earth, after cellulose.

1.3 Structural Colours in Other Scale-Bearing Insects

Other scale-bearing insects are known to display structural colours (Ghiradella, 1998). Species from the Scarabaeidae family (scarab beetles), the Curculionoidea superfamily (weevil beetles), the Cerambycidae family (longhorn beetles) and the Formicidae family (ants) are examples. Scales on insect bodies vary in shape and size. When specifically elongated, they are termed bristles and sometimes hairs (Ghiradella, 1998) and can also be the seats of colour. Examples of such systems now follow.

1.3.1 Hoplia coerulea *Scarab Beetle*

The blue-violet iridescence of the male *Hoplia coerulea* beetle is the result of a photonic structure (a periodic multilayer—see Section 3.2) located in the round scales covering its body (Fig. 2) (Mouchet et al., 2017a; Vigneron et al., 2005a). Interestingly, these scales were shown to exhibit a change in colour upon contact with various liquids and vapours following their penetration into the scale bodies and the filling of air pores in their structure (Fig. 2E) (Mouchet et al., 2016b,c, 2017b; Rassart et al., 2009). Additionally, fluorophores were found to be embedded within the scales and to emit a turquoise colour under UV illumination (Mouchet et al., 2016a;

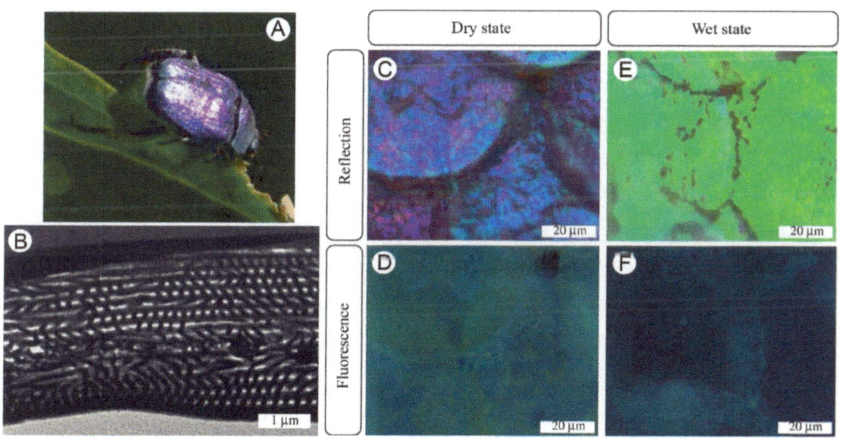

Fig. 2 The male *H. coerulea* beetle exhibits *blue-violet* iridescent elytra and thorax (A) due to a periodic multilayer (B) incorporated in the scales covering its cuticle (C). Illuminated by UV light, the scales appear *turquoise* due to fluorescence emission (D). Upon contact with water, the scales turn to *green* (E) under visible white light and *navy blue* (F) under UV light. *Reproduced from Mouchet, S.R., Lobet, M., Kolaric, B., Kaczmarek, A.M., Van Deun, R., Vukusic, P., Deparis, O., Van Hooijdonk, E., 2016. Controlled fluorescence in a beetle's photonic structure and its sensitivity to environmentally induced changes. Proc. R. Soc. B 283, 20162334.*

Van Hooijdonk et al., 2012b) that becomes navy blue after the deposition of a water droplet on the beetle elytron (Fig. 2D and F) (Mouchet et al., 2016a, 2017a).

1.3.2 Weevil and Longhorn Beetles

The scales covering the elytra and bodies of many weevil and longhorn beetles also comprise photonic structures. They are often more or less elongated, sometimes taking even the shape of bristles. The nanostructures are generally either very ordered and periodic in three dimensions (referred to generally as a three-dimensional (3D) PC—see Section 3.4) or randomly disordered. The juxtaposition of different domains (also termed grains) of such crystals, the size of which can reach 2000 μm^2, within single scales and with different orientations can lead to multilength-scale visual effects such as additive colour mixing (Mouchet et al., 2012b, 2013). Such a set of PC domains are often referred to as photonic polycrystals.

One example of this is the scales covering the dark elytra of *Entimus imperialis* weevil (Deparis and Vigneron, 2010; Mouchet et al., 2012b, 2013; Wilts et al., 2012a,b; Wu et al., 2013). The matt angle-independent green patches observed on these elytra are due to sets of scales, the photonic domains of which comprise the same photonic structure with different orientations and display colours ranging from orange to blue (Fig. 3). This appearance is believed to improve the insect camouflage. Similar photonic nanostructures were found in the scales covering the elytra of *Pachyrrhynchus congestus pavonius* (Welch et al., 2007), *Lamprocyphus augustus* (Galusha et al., 2008), *Eupholus cuvieri* (Saranathan et al., 2010), *Eupholus magnificus* (Pouya et al., 2011) weevil beetles and *Prosopocera lactator* longhorn beetle (Colomer et al., 2012). Noteworthy, a periodic multilayer similar to the one found in *H. coerulea*'s scales was identified in the scales incorporated on the elytra of *Tmesisternus isabellae* longhorn beetle (Liu et al., 2009).

Randomly disordered structures were found in the scales on the elytra of some longhorn beetles. Light interacting with such structures (see Section 3.5) gives rise to a white colouration such as in the case of *Calothyrza margaritifera* (Lafait et al., 2010). Similar structures and effects were previously identified in the scales covering *Cyphochilus* spp. and *Lepidiota stigma* beetles from the Scarabaeidae family (Burresi et al., 2014; Vukusic et al., 2007). *Celosterna pollinosa sulfurea* and *Phosphorus virescens* beetles combine such disordered structures to yellow and fluorescent pigments in order to give rise to their yellowish appearances (Van Hooijdonk et al., 2013).

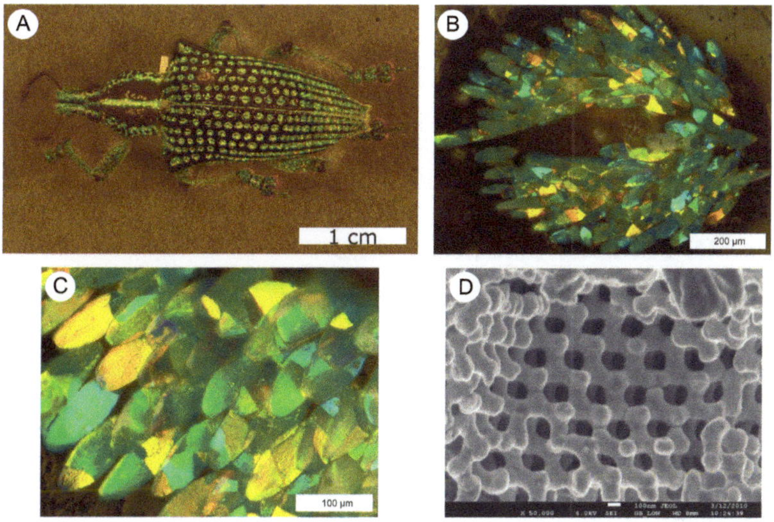

Fig. 3 The matt *green spots* observed on the elytra of the Brazilian weevil *E. imperialis* (A) are due to scales gathered around pits and exhibiting colours ranging from *blue* to *orange* (B, C). The close juxtaposition of the scales comprising PC domains with different orientations of the same gyroid photonic structure (D) gives rise to additive colour mixing. *Reproduced from Mouchet, S.R., Vigneron, J.-P., Colomer, J.-F., Vandenbem, C., Deparis, O., 2012. Additive photonic colors in the Brazilian diamond weevil,* Entimus imperialis. *Proc. SPIE 8480, 848003.*

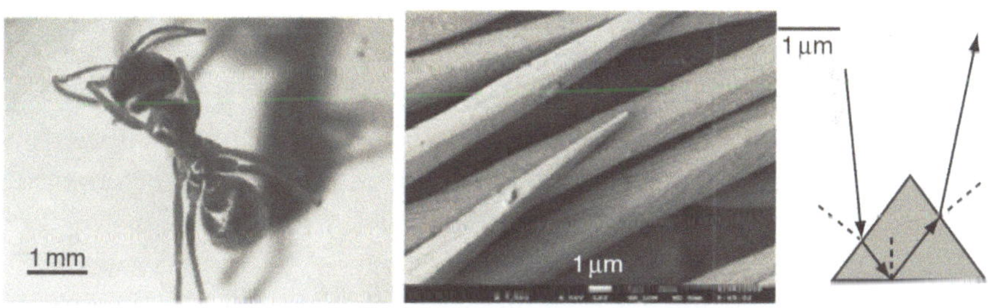

Fig. 4 The body of *C. bombycina* ants is covered with bristles giving rise to a *silver appearance*. The *triangular geometry* of the cross-section of these bristles leads to total internal reflection of incident light. *Reproduced from Vigneron, J.-P., Simonis, P., 2010. Structural colours. Adv. Insect Physiol. 38, 181–218, with permission of Elsevier.*

1.3.3 Cataglyphis bombycina *Ant*

The heads, the thoraxes and the abdomens of *Cataglyphis bombycina* ants have a silver appearance. They are covered by bristles (Fig. 4), the cross-section of which is an equilateral triangle with about 5 µm sides (Vigneron and Simonis, 2010). The base of each bristle is flat and parallel to the ant's cuticle surface. Instead of giving rise to light diffusion, these transparent bristles lead to total internal reflection at the basal interface of the bristles (Fig. 4), due to their morphology (Shi et al., 2015; Vigneron and Simonis, 2010). This high reflection allows this ant species to tolerate high temperature in its native Sahara desert.

2. CLASSIFICATION OF PHOTONIC NANOSTRUCTURES IN LEPIDOPTERAN SCALES

Structural colours in butterfly wings are usually due to ordered and nanostructured scales covering their wings. Formed in the chrysalis (Ghiradella, 1989), these scales, with typical dimensions of $200 \times 75 \times 2\,\mu m^3$, are flattened sac of cuticle, regularly imbricated in rows and fixed to the wing membranes with flexible stalks plugged into sockets. Scales are usually found on each side of a wing, in two layers, referred to as ground scales and covering scales.

Two main layers, also called laminae, comprise in a single butterfly wing scale (Ghiradella, 1998; Vukusic et al., 2000a). The basal lamina is flat and solid. The upper lamina is structured. Both laminae are connected by pillar-shaped trabeculae. The general template of all butterfly scale designs (Fig. 5), whether iridescent or not, presents longitudinal ridges in the upper lamina, parallel to the main axis of the scale and extending from the stalk to the top. These ridges are usually uniformly spaced with a pitch within the range 500–5000 nm and connected to each other by arches called crossribs, forming windows to the scale body. They constitute a layer or are laid on a membrane. On the ridges, finer nanostructures can be found in the form of fluted microribs or lamellae. In most scales, the spaces between the ridge layer and the scale lower lamina beneath are filled with air. Other structures, sometimes periodic, can be found in these spaces. Melanin may be distributed in all the scale material. Other pigments, if present, are usually found in granules in these windows between ridges and crossribs.

Lepidopteran scale classification can be based on the morphological elements of the structure reflecting, diffracting or scattering incident light as well as on their location on the scales (Vukusic, 2005; Vukusic et al., 2000a). Three main categories—type I, II and III—can be singled out (Fig. 5). Each of them can be further divided into subcategories. All these types are variations of the generic design. Type I scales, also called ridge-lamellae scales, display discrete multilayer structures on the ridges. Type II scales, known as body-lamellae scales, exhibit multilayers within the body of the scale. Type III scales, termed body-scattering scales, present other

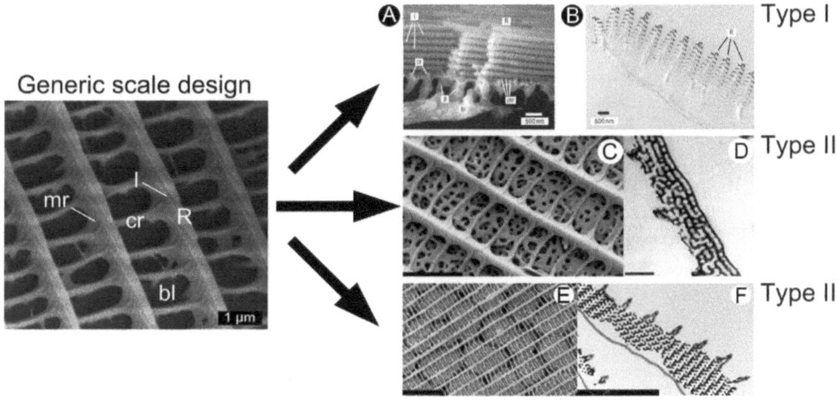

Fig. 5 Classification of the nanostructures of iridescent butterfly scales. The generic scale design exhibits parallel longitudinal ridges (R) connected by crossribs (cr) in the upper lamina, as observed here in *Apatura ilia* ground scales. Microribs (mr) are also incorporated in these ridges. Pillars (p) connect the upper lamina to the basal lamina (bl). Type I scales display ridges incorporating multilayers, as observed in *Morpho sulkowskyi* scales by SEM (A) and TEM (B). Type II scales comprise continuous multilayer structures within their body, as found in *Albulina metallica* dorsal wing scales (C, D). The scale bars in (C) and (D) correspond to 2 and 1 µm, respectively. Complex light scattering photonic structures are incorporated in type III scale body such as in the ventral wing scales of *Cyanophrys remus* (E, F). Both scale bars (E, F) correspond to 5 µm. *Panels (A) and (B): Reproduced from Potyrailo, R.A., Ghiradella, H., Vertiatchikh, A., Dovidenko, K., Cournoyer, J.R., Olson, E., 2007.* Morpho *butterfly wing scales demonstrate highly selective vapour response. Nat. Photonics 1, 123–128, with permission of Nature Publishing Group. Panels (C)–(F): Reproduced from Biró, L.P., Kertész, K., Vértesy, Z., Márk, G.I., Bálint, Z., Lousse, V., Vigneron, J.-P., 2007. Living photonic crystals: butterfly scales—nanostructure and optical properties. Mater. Sci. Eng. C 27, 941–946, with permission of Elsevier.*

nanostructures such as 3D PCs and randomly disordered structures in the scale bodies.

All these lepidopteran photonic structures are generally open to air, the surrounding environment. This is, in contrast to scales covering, the elytra of beetles for instance (Section 1.3). This feature makes them able to exchange fluids with the surrounding environment since these fluids can easily penetrate their open structures. Upon contact with liquids of low surface tension such as ethanol (Fig. 6), structurally coloured butterfly wing scales are known to change their colour appearances (Vukusic et al., 1999; Wang et al., 2014). This form of fluid penetration test enables rapid differentiation between structural and pigmentary colours in such systems, the liquid application leading to suppressing of strong coherent scattering through an RI-matching mechanism. Colour changes induced in such open structurally

Fig. 6 The *blue colour* of *M. sulkowskyi* butterfly (as observed here on the hindwing) turns to *green* (as displayed by the forewing) upon intrascale penetration with isopropanol.

coloured systems as a result of contact with vapours and gases were also observed with butterfly wings (Biró et al., 2008; Kertész et al., 2013; Mouchet et al., 2012a; Piszter et al., 2014; Potyrailo et al., 2007, 2013). Any biological function associated with these colour changes is, however, so far undetermined. Colour changes induced by variations in temperature (from ambient temperatures down to −10°C) were reported in several species (Kertész et al., 2013, 2014; Tamáska et al., 2013). These colour changes induced by contact with vapours and temperature variations have been mainly explained by vapour adsorption in the structure pores (Mouchet et al., 2012a; Piszter et al., 2014; Potyrailo et al., 2013). One of the technological applications of this phenomenon is the development of gas and vapour sensors, through a bioinspiration approach (Gao et al., 2011; Jiang et al., 2014; Poncelet et al., 2016; Potyrailo and Naik, 2013; Rasson et al., 2017; Yang et al., 2011). Additionally, real butterfly wings scales were modified in order to exhibit colour changes induced by variations of temperature and pH using, respectively, mono-wall carbon nanotubes, increasing thermal coupling between incident electromagnetic waves and the nanostructure (Pris et al., 2012), and a pH-sensitive polymer that swells or contracts as a function of ambient pH (Zang et al., 2012).

2.1 Type I Scales

Many species of Lepidoptera possess scales that have been referred to in previous classifications as type I scales (Fig. 5) (Vukusic, 2005; Vukusic et al., 2000a): these scales are classified as exhibiting ridges that incorporate colour-generating multilayering. Colours generated by such structures arise due to light interference in these multilayers (Section 3.2). Subcategories of type I have been described and these are related to the angle formed between the ridge multilayering and the underlying wing scale substrate.

Type Ia scales are classified as comprising relatively periodic lamellae that run parallel or almost parallel to the scale base lamina. The archetype of this subcategory is the scales found on the wings of the species from the *Morpho* genus, by far, the most investigated case of butterfly photonic structures (Berthier, 2003, 2010; Berthier et al., 2006; Kinoshita et al., 2002a,b; Vukusic et al., 1999; Yoshioka and Kinoshita, 2004). The photonic structures of *Morpho*-type scales are often compared to Christmas trees because of the shapes of the cross-sections of their spatially discrete multilayers observed by TEM. They most often exhibit the brightest lepidopteran scales with striking blue iridescent metallic colour appearance.

The second subcategory, type Ib, is described as comprising scales bearing lamellae on their ridges that are inclined at a moderate angle with respect to the wing scale lamina. As a result of the inclination, this form of scale structure can give rise to a larger range of colours as exemplified in the *Ancyluris meliboeus* species (Vukusic et al., 2001b, 2002). The inclination also restricts the solid angle over which incident light is reflected, creating a strong on–off colour flashing facility with minimal wing movement.

Scales with a very large angle of inclination between the lamellae and the scale base belong to type Ic subcategory. This angle can in some species be close to the normal to the scale lamina. The solid angle, over which light reflected from the scales, is very small, as is the case in *Troides magellanus* for which bright structurally coloured reflection is only visible from the species' rear at grazing incidence.

2.2 Type II Scales

Continuing the classification exercise, structurally coloured butterfly wing scales displaying flat or curved continuous multilayer structures within their body have been described as type II scale structures (Fig. 5). These multilayers alternate high RI layers and low RI air spaces between them. The former are usually dense cuticle layers that can exhibit so-called "pepper-pot" perforations (Biró and Vigneron, 2011; Kinoshita, 2008). The latter are typically air layers comprising often a fibrous network supporting the dense layers. Up to 10 of these layers are typically found in such systems. Like type I scales, the resulting structural colours arise due to multilayer interference (Section 3.2). In most species exhibiting this form of scale structure such as *Urania leilus* moths (Vukusic and Sambles, 2001), these scales also have ridges on their surface that have a less elaborated profile that do not seem to play a significant role in the scales' optical signature. This category is divided into two subcategories depending on the curvature of the multilayer.

Type IIa scales are characterised by lamellae parallel to the scale lamina and are relatively flat across the whole scale. Crossribs usually connect periodically the ridges over the multilayer nanostructure.

In contrast with type IIa scales, the lamellae on type IIb scales present a modulation that is often curved along the direction normal to the layers. *Papilio palinurus* and *Papilio ulysses* are species exhibiting this form of structurally coloured wing scale (Vukusic et al., 2000b, 2001a).

2.3 Type III Scales

Structural colours arising from, what has been classified as type III scales, are not created by interference multilayers such as in type I and II scales. Instead, this colour arises by light scattering in more complex photonic structures within the body of the scale (Fig. 5).

Two- or three-dimensional PC structures are found in the body of these type IIIa scales (Section 3.4), lying between the ridges and the scale base lamina, such as in the case of *Callophrys rubi* (Allyn and Downey, 1976; Michielsen and Stavenga, 2008; Michielsen et al., 2010) and *Parides sesostris* (Michielsen and Stavenga, 2008; Prum et al., 2006; Saranathan et al., 2010; Vukusic and Sambles, 2001). These PCs are usually formed of distinct irregular domains that exhibit different orientations within a single scale. The domains, the size of which is typically a few microns, can lead to multilength-scale colours in these photonic polycrystal structures (Kertész et al., 2006; Morris, 1975) where incident light is scattered differentially by different neighbouring domains. The visual ensemble appearance of these photonic polycrystals, however, is created by additive colour mixing across the domains, leading to angle-independent and hence noniridescent colour.

In type IIIb scales such as found on *Pieris rapae* and *Delias nigrina* wings (Stavenga et al., 2004, 2006), randomly distributed structures (Section 3.5) are found in the scale body between the ridge basis and the scale basal lamina, instead of periodic nanostructures, leading to light incoherent scattering.

3. PHYSICAL ORIGINS OF COLOURS IN BUTTERFLY WING SCALES

The interaction of light with inhomogeneous structured media gives rise to coherent and incoherent scattering, depending on whether the media are geometrically ordered or disordered. Coherent scattering arises from interference in ordered systems such as in optical thin films, multilayers, diffraction gratings and PCs. All these kinds of systems can be found in many animals and many plants. In order to achieve specific visual appearance, many organisms combine different sorts of optical systems, leading sometimes to multilength-scale optical effects. Examples of this will be presented in Section 4.

Among ordered optical media, PC systems can be the most visually striking. These structures comprise periodic variations of the RI (Joannopoulos et al., 2008; John, 1987; Yablonovitch, 1987). In general, they are classified into three categories, according to the dimensionality of their periodicity: one-dimensional (1D), two-dimensional (2D) and three-dimensional (3D). These periodicities influence light propagation. As a result of constructive and destructive interference within the structure, light propagates or is inhibited from propagating, depending on its wavelength. A wavelength that gives rise to light propagation is called a mode and a set of propagating modes are termed a band. A range of wavelengths within which light is forbidden to propagate is a gap in propagation bands and is referred to as a photonic bandgap (PBG). This model is analogous to concepts in solid-state physics associated with electrons propagating in solids. By way of representing both the electronic and optical models, electronic band structures and photonic band structures can be plotted for atomic and for PCs, respectively. The colour reflected from a system is associated with a band of light waves that are forbidden from propagating through the system which, accordingly, are then scattered back in the incident direction. Periodic modulations of the RI in the direction normal to the interface between the incidence medium and the PC, as well as in the lateral directions, give rise to specular reflection and potentially also to diffraction, respectively. In contrast to most pigmentary colours, light scattering or reflection by these sorts of nanostructures, naturally occurring in insect systems or synthetically produced by lab-based manufacture, is often directional and the associated colours can be very vivid. For nonoptically absorbing materials, PCs that comprise a very high number of repeating structural units, or periods, and an appropriate combination of different RI materials can be perfect reflectors of bands of incident colour. Such bands of colour, corresponding to the system's bandgaps, if they exist in every orientational direction and for any light polarisation, are called complete PBGs. Often though, they do not arise in natural photonic structures due to the insufficient contrast between the component RI. The RI of air is commonly assumed to be equal to $n_{air}=1.00$ in the visible range (Birch and Downs, 1993, 1994; Edlén, 1966). The RI of lepidopteran cuticle, often supposed principally to comprise chitin, has been measured to be approximately $n_{cut}=1.56$ (Mason, 1926; Sollas, 1907; Vukusic et al., 1999; Yoshioka and Kinoshita, 2011).

A perfect PC is somewhat of an idealised structure that does not exist in real life. Perfect PCs should be infinitely periodic and comprise distinctly different RI materials. In living organisms, PC nanostructures are neither spatially infinite nor perfectly ordered. To this end, they are sometimes described as pseudo-ordered and are therefore often referred to as being PBG materials instead of PCs. Pseudo-order (called sometimes quasi-order), disorder or general variations at nanometric scale can be found in the physical geometry or in the RI of the constituent materials (Boulenguez et al., 2012; Vigneron et al., 2005b): this can arise, for instance, through the presence of irregular interfaces or variations in RI. Bright colour reflection is of lower intensity as a result of the presence of such defects. Peaks in reflectance spectra may be broader (leading to less saturated colour appearances) or can be less intense or the solid angles over which scattering occurs may be wider. Micrometric-scale variations in structural order can also be identified: examples of this include scales covering some butterflies and beetles that are misaligned (Boulenguez et al., 2012; Vigneron et al., 2005b) or intrascale PC domains that present different orientations (Kertész et al., 2006; Mouchet et al., 2012b; Vigneron et al., 2005b; Vignolini et al., 2012). These sorts of defects may give rise to multilength-scale optical effects that can in some cases lead to additive colour mixing. In most photonics technological applications, disorder and defects can be critical. In natural system, however, photonics structures and their optical responses usually appear very tolerant of, or resilient to, structural or material defects and imperfections. Despite significant imperfections, the visual appearance of a species, and the inherent optical functions of that appearance, evolved for over millions of years, is reproduced effectively. From a bioinspired technology perspective, the biological photonic structures present in many of the insect systems described here and across the literature, the optical responses of which appear resilient to defects and gross imperfections, are valuable and may form the basis of future applications and devices.

3.1 Interference in Thin Films

Interference in a single thin film is a common origin of structural colour. To understand the process by which this colour is produced, consider a thin film with RI n_2 and thickness t between an incident medium with an RI equal to n_1 and a substrate of RI n_3. Incident light with a wavelength λ, forming an angle θ with the normal to the film, will reflect at both interfaces of the film: these two reflections will interfere with each other (Fig. 7A). The optical path difference between rays reflected from the top and the bottom interfaces is $2dn_2 \cos\theta$. The condition for constructive interference between these two reflections will depend on the RI of the substrate. When incident light reflects on an interface with an emergence medium with a higher RI, the phase of the light wave is modified by 180° (Hecht, 1998; Yeh, 2005). This leads to constructive interference conditions:

$$\left(m - \frac{1}{2}\right)\lambda = 2dn_2 \cos\theta, \quad \text{for} \quad n_1 < n_2 \text{ and } n_3 < n_2, \tag{3}$$

as well as

$$m\lambda = 2dn_2 \cos\theta, \quad \text{for} \quad n_1 < n_2 < n_3, \tag{4}$$

where m is an integer. Such interference occurs in a soap bubble (Eq. 3) and glass antireflective coatings (Eq. 4), respectively. These conditions do not take however multiple reflections at the interfaces into account. Exact calculation can be derived by summing all reflected ray amplitudes (Fig. 7B) (Yeh, 2005).

3.2 Interference in Periodic Multilayers

A 1D PC is the simplest case of PCs. It consists of a periodic multilayer comprising at least two types of homogeneous layers of alternating different RI and of thicknesses that may not be equal (Fig. 8). Such a photonic structure is also referred to as a "Bragg mirror" or a "Bragg reflector" and plays the role of an optical filter, selectively reflecting specific wavelengths of incident light in the specular direction. It is very common not only in scale-bearing Lepidoptera but also in other fauna and flora (Kinoshita, 2008; Land, 1972).

A 1D multilayer comprising a large number of layers and strong RI contrast between the constituent layer materials can exhibit an intense and narrow-band reflectance spectrum, the reflectance peak wavelength or colour of which depends on the incidence and observation angles (Fig. 8). This

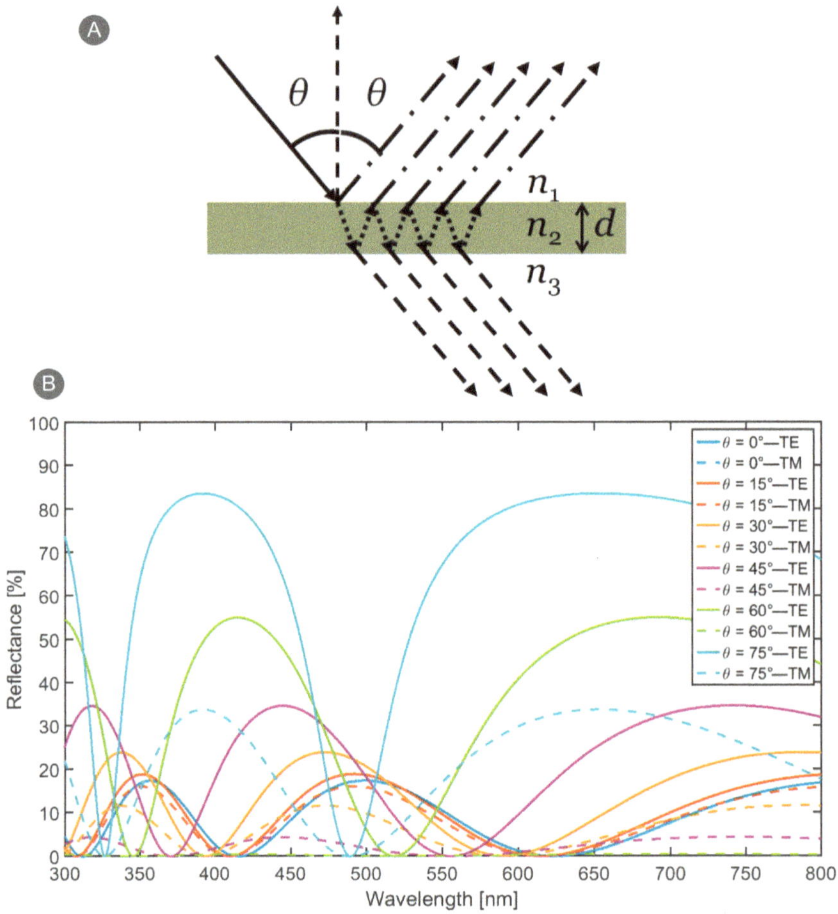

Fig. 7 (A) Thin-film interference in a film with an RI n_2 and a thickness *t* between two semi-infinite media with RI equal to n_1 and n_3. The incidence angle with respect to the normal to the film interfaces is θ. (B) Reflectance spectra calculated for a 400-nm thick film with an RI $n_2 = 1.56$ between two media with RI equal to $n_1 = n_3 = 1.00$ for different values of θ. The spectra for both transverse electric (TE) and transverse magnetic (TM) light polarisations depend on the incidence angle. At normal incidence ($\theta = 0^{\circ}$), TE and TM curves are identical.

angle-dependent reflectance is the origin of the iridescent appearance of the layered system.

From an historical point of view, one of the first mathematical treatments of iridescent reflection from a periodic multilayer arose from Lord Rayleigh's treatment of the optical properties of thin film in 1887 (Rayleigh, 1887, 1888).

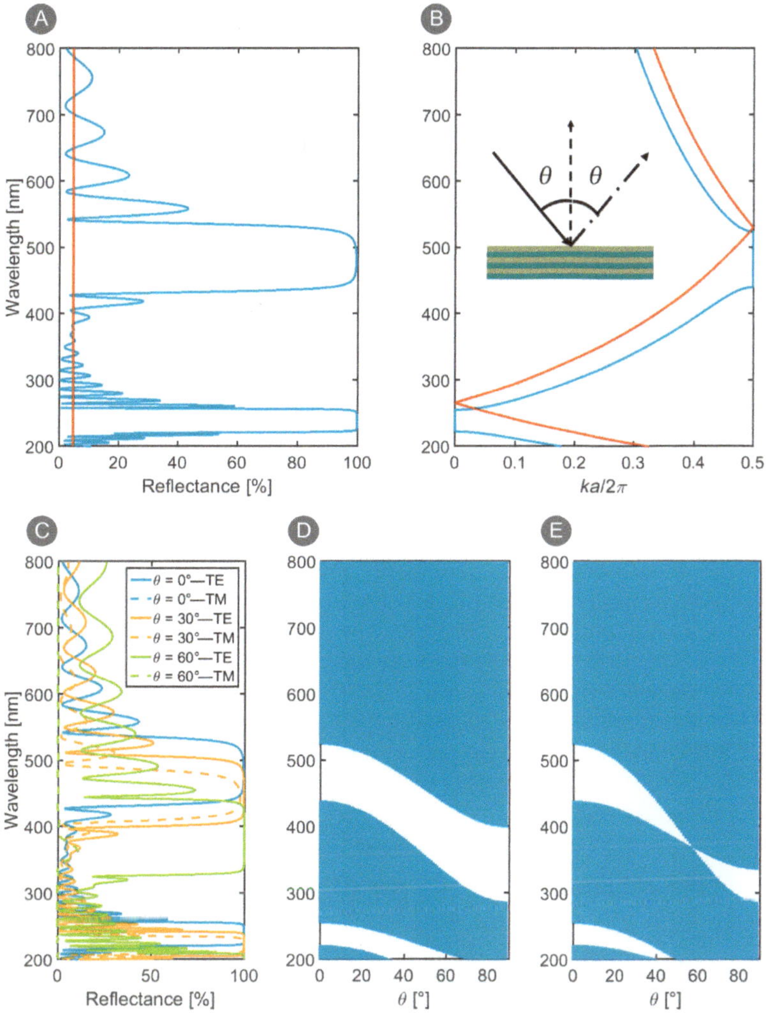

Fig. 8 (A) Reflectance spectra calculated at normal incidence ($\theta=0^\circ$) from a periodic multilayer (*inset* in B) comprising 12 bilayers, the thicknesses of which are equal to $d_1=120$ nm and $d_2=50$ nm and RI $n_1=1.56$ and $n_2=1.00$ (*blue curve*) between two semi-infinite media, the RI of which are $n_{inc}=1.00$ and $n_{sub}=1.56$; and from an interface between these two semi-infinite media without multilayer (*red curve*). The reflectance spectrum from the multilayer exhibits a peak (more than 99%) located at approximately 480 nm. (B) Photonic band structure related to an infinite periodic multilayer comprising the same two layers as in (A) (*blue curve*) and to an infinite periodic multilayer comprising layers with the same thicknesses as in (A) and with RI equal to $n_1=1.56$ and $n_2=1.56$ (*red curve*), i.e., without RI contrast. The RI contrast leads to the creation of the PBGs. The shift in bandgap positions is due to the difference in effective RI. (C) Reflectance spectra calculated from the same periodic multilayer as in (A), for both transverse electric (TE) and transverse magnetic (TM) polarisations and different incidence angles θ. At normal incidence ($\theta=0^\circ$), TE and TM spectra are identical. (D and E) Related photonic band structures as a function of the incidence angle for TE (D) and TM (E) polarisations. *Blue areas* correspond to allowed photonic bands.

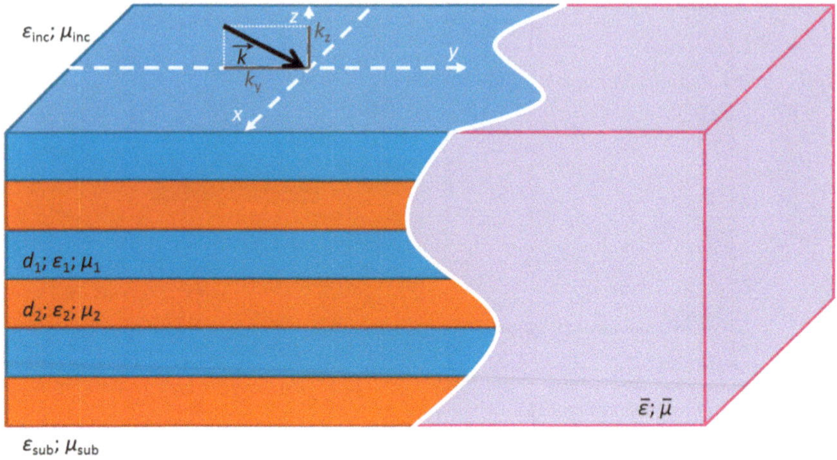

Fig. 9 Wavevector $\vec{k}$ penetrating a multilayer comprising a periodic stack of two different layers with respective thicknesses equal to d_1 and d_2, dielectric permittivities equal to ε_1 and ε_2 as well as magnetic permeabilities equal to μ_1 and μ_2. The dielectric permittivities of the incidence medium and the substrate are, respectively, ε_{inc} and ε_{sub} and their magnetic permeabilities are, respectively, μ_{inc} and μ_{sub}. Such a multilayer can be approximated by a homogeneous medium, the dielectric permittivity and magnetic permeability of which are $\bar{\varepsilon}$ and $\bar{\mu}$, respectively (*right*).

The peak reflectance wavelength, or dominant wavelength, related to a large number of periodic layers in a multilayer system can be approximately calculated with the *dominant reflected wavelength* formula. If a 1D multilayer system comprises two layers of thicknesses d_1 and d_2, and, dielectric permittivities $\varepsilon_1 = n_1^2$ and $\varepsilon_2 = n_2^2$, respectively, where n_1 and n_2 are the RI of the related layers (Fig. 9), then the dominant reflected wavelength is given by (Deparis et al., 2006; Vigneron and Lousse, 2006; Welch et al., 2007):

$$\lambda_{\text{dom}} = \frac{2a\sqrt{\bar{n}^2 - n_{\text{inc}}^2 \sin^2\theta}}{m}. \tag{5}$$

Here $a = d_1 + d_2$ is the multilayer period; $\bar{n}$ is the effective RI; n_{inc} corresponds to the RI of the incident medium; θ is the incident angle; and m is a natural number chosen so that λ_{dom} belongs to the range of interest (in our case, visible range). Depending on the value of m, λ_{dom} corresponds to the fundamental reflection wavelength ($m = 1$) or the harmonics ($m > 1$) of constructive interference Bragg reflection from the system. This formula also predicts iridescent behaviour of the system; namely, with larger incidence angle θ, the dominant wavelength λ_{dom} decreases. This shift of the peak reflectance wavelength is referred to as *blueshift*. The inverse shift is a *redshift*.

The effective RI $\overline{n}$ can be calculated by an effective medium approximation using, for instance, the Maxwell Garnett equation or, in a much simpler way, by using the approximation formula:

$$\overline{n} = \sqrt{\frac{\varepsilon_{\parallel} + \varepsilon_{\perp}}{2}}, \tag{6}$$

where $\varepsilon_{\parallel} = (d_1/a)\varepsilon_1 + (d_2/a)\varepsilon_2$ and $\varepsilon_{\perp}^{-1} = (d_1/a)\varepsilon_1^{-1} + (d_2/a)\varepsilon_2^{-1}$. This approximation is valid for small RI contrasts and with incident wavelengths large with respect to the multilayer period.

The optical response of such a structure can be spectrally simulated by thin-film solver software using formalisms, which in addition take light polarisation into account, such as transfer matrix methods, recursion techniques or iterative approaches (Hecht, 1998; Huxley, 1968; Yeh, 2005).

3.3 Diffraction by Gratings

1D diffraction gratings that generate colour reflection are less common in lepidopteran systems. They have been, however, observed in a few species (Allyn and Downey, 1976; Ingram et al., 2008; Vigneron et al., 2010; Vukusic et al., 2002, 2009). 1D grating structures most usually have a surface which displays a periodicity along the surface (rather than under it as is the case with multilayering). This divides incident broadband light (white light for instance) into several diffracted (reflected or transmitted) orders in different directions. This leads to an angular distribution of scattered colours, visually appearing like the reflection from the silvered surface of a CD, DVD or Blu-ray Disc (Fig. 10). 1D diffraction gratings reflecting discernible colour signals often have their repeating periodic structure in the size of 1 to several micrometres.

In considering a 1D diffraction grating with a periodicity d that is illuminated with broadband incident light at an angle θ_{inc}, it can be shown that the beam diffracted with an angle θ_{dif} will be characterised by a wavelength equal to:

$$\lambda_{dif} = \frac{d(\sin\theta_{inc} + \sin\theta_{dif})}{m}, \tag{7}$$

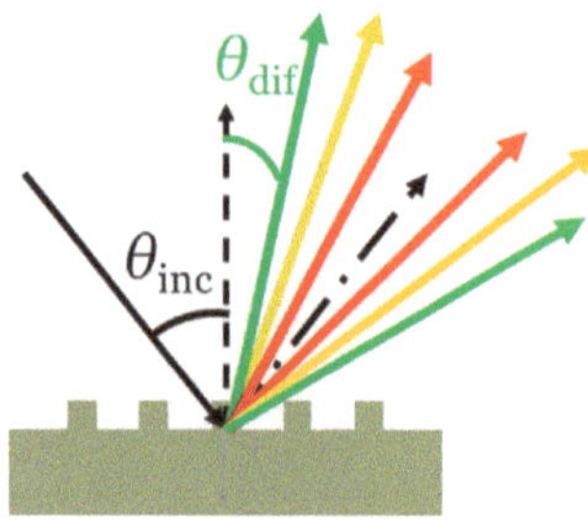

Fig. 10 The lateral periodicity of a diffraction grating disperses incident white light into diffracted orders giving rise to a wavelength-dependent angular distribution of light.

where m is an integer that represents the diffraction order. In contrast to the iridescent reflection from layers, Eq. (7) indicates that longer wavelengths will be diffracted at larger angles.

3.4 Light Interaction With 3D Photonic Crystals

3D PCs present a periodicity of RI variation along the three dimensions of space. These sorts of structures have been observed in the scales of a number of butterfly species (Kertész et al., 2006; Michielsen and Stavenga, 2008; Pouya and Vukusic, 2012; Vukusic and Sambles, 2001).

The concept of a PC was coined after the discovery made by Yablonovitch (1987) and John (1987). Their studies showed the existence of a PBG for 2D and 3D periodic structures. More recently, PC systems are used in many devices and for a range of applications: PC fibres (Foresi et al., 1997), light-emitting diodes (Boroditsky et al., 1999; Erchak et al., 2001), photocatalysis (Chen et al., 2006, 2008; Deparis et al., 2015) and light trapping (Demésy and John, 2012).

The optical properties of such photonic structures can be predicted using thin-film solver software (Mouchet et al., 2013) as well as the dominant wavelength formula (Kertész et al., 2006; Vigneron and Lousse, 2006; Welch et al., 2007), in a similar way to multilayered systems. This requires the approximation of, for instance, a 3D PC structure by a 1D photonic structure. In such cases, the photonic structure with which the incident beam interacts along a specific crystallographic direction can be associated with a periodic stack of homogeneous layers lying perpendicular to the selected crystallographic direction. The peak reflectance will be located at:

$$\lambda_{dom} = \frac{2p\sqrt{\bar{n}^2 - n_{inc}^2 \sin^2\theta}}{m}, \tag{8}$$

where p is the distance between two reticular planes of the system at the specific crystal orientation being considered: this replaces the period a in Eq. (5). A family of reticular planes is a set of parallel planes including all the nodes of the lattice and so that each plane includes at least three nodes. The weighting factors in averages performed for the calculation of the effective RI (Eq. 6) are $f = V_1/V_c$ and $1-f$, where V_1 is the volume of material, the dielectric permittivity of which is ε_1, within the crystal cell and V_c is the volume of the crystal cell.

As shown in the previous section, when an incident beam interacts with a surface that exhibits grating-like structural periodicity, light may diffract from it. In the simplified case of normal incidence $\theta_{inc} = 0^\circ$, the dominant wavelength of each diffracted order g is calculated using (Rassart et al., 2009):

$$\lambda_g = \frac{2p\bar{n}}{\sqrt{m^2 + \left(\frac{pg}{\pi}\right)^2}}. \quad (9)$$

Rigorous computations of the optical properties of 3D PCs are usually performed using formalisms such as the Finite-Difference Time-Domain (FDTD) method (Taflove and Hagness, 2005; Yee, 1966), the Finite Element Method (FEM) (Davies, 1980; Dhatt and Touzot, 1984), the Rigorous Coupled-Wave Analysis (RCWA) method (Moharam and Gaylord, 1981; Vigneron and Lousse, 2006) or a frequency-domain eigensolver of the Maxwell eigenproblem (Joannopoulos et al., 2008; Johnson and Joannopoulos, 2001). Series of commercial software comprising these approaches is available.

3.5 Light Incoherent Scattering in Randomly Disordered Structures

Light interaction with randomly disordered structures is known to give rise to incoherent scattering processes. Particles, referred to as scatterers, the sizes of which are on the order of the wavelength of interest (visible light in our case), scatter incident light in all directions regardless of the incident wavelength. Such optical effects have been observed in a few butterfly species (Morehouse et al., 2007; Stavenga et al., 2004, 2006). The displayed colour appearances are noniridescent. If the scatterers have a spherical shape, the optical response can be predicted on the basis of Mie theory (Born and Wolf, 1975; Hecht, 1998; Mie, 1908). With nonspherical scatterers, no exact solution can be found analytically and the light scattering may be described by T-matrix method (Mishchenko et al., 1996).

It was previously believed that incoherent scattering was responsible for the blue wings' colour appearance of the male *Papilio zalmoxis* butterfly (Huxley, 1976): this process was even referred to as Tyndall scattering. More recent studies on *P. zalmoxis* and other *nireus* group species (Trzeciak et al., 2012; Vukusic and Hooper, 2005; Wilts et al., 2012c) have shown that their wing scale structures do not give rise to incoherent scattering: instead, a combination of fluorescence and structural colour effects is the cause.

4. CASE STUDIES IN LEPIDOPTERAN SCALE STRUCTURES

In this section, characteristic examples of each scale category described in Section 2 are reviewed, from type I to type III scale species.

In addition to the related optical effects, natural photonic structures in butterfly wings exhibit supplementary properties such as colour changes induced upon contact with fluids (Biró et al., 2008; Mouchet et al., 2012a; Potyrailo et al., 2007, 2013; Vukusic et al., 1999), light trapping (Herman et al., 2011; Vukusic et al., 2004) and strong polarisation signatures (Vukusic et al., 2000b, 2001a; Yoshioka and Kinoshita, 2007). Structures that exhibit several of these features concurrently are clearly multifunctional with respect to their physical and optical properties. However, it is not always clear whether these properties are related to biological functions.

Although most lepidopteran wings are naturally covered dorsally and ventrally in scales, several species such as *Cacostatia ossa* moths, *Cephonodes hylas* moths and butterflies from the *Greta* genus exhibit optical effects arising from photonic structures without any scales covering their wings (Deparis et al., 2009, 2014; Siddique et al., 2015; Stavenga, 2014; Yoshida, 2002; Yoshida et al., 1996, 1997). These species exhibit very transparent wings with surface nanostructures that reduce light reflection from the wings, in this way serving an antireflection function. These highly transparent wings may play a key camouflage role.

4.1 *Morpho* Genus

Butterflies from the *Morpho* genus are Neotropical butterflies originating from Central and South America. The colour appearance of the dorsal surfaces of their wings varies between species, from metallic blue, to diffuse blue, to white and even brown-yellow (Berthier, 2010; Berthier et al.,

2006). The ventral faces of their wings are usually patterned and generally brown, this brown colour arising from the presence of melanin pigmentation.

The iridescent blue appearance of species, such as *Morpho rhetenor*, *Morpho didius*, *Morpho menelaus* and *M. sulkowskyi*, has attracted a lot of attention from entomologists, butterfly collectors and researchers in photonics for many years. Such bright colours are known to play a role in long-range intraspecific communication and courtship (Silberglied, 1984). They present a sexual dimorphism. The males of some species such as *M. rhetenor* exhibit bright blue appearances on the dorsal faces of their wings, whereas *M. rhetenor* females exhibit dull yellowish and brownish colours. In cases like *M. didius*, both sexes have an iridescent blue colour but females are less bright (Berthier, 2010; Berthier et al., 2006).

The iridescent blue *Morpho* butterflies generally exhibit type Ia scales (Fig. 11). The lamellae incorporated into the ridges of these scales form discrete multilayers comprising alternating air and cuticle layers with as many as 12 cuticle layers in some species such as *M. rhetenor* (Vukusic et al., 1999). The main origin of the exhibited colouration is constructive and destructive interference within the multilayer nanostructures giving rise to a wavelength-selective reflection. At short visible wavelengths, reflectance from *M. rhetenor*'s wings can reach more than 75%, due to the high number of layers in the multilayering as well as the very small separation between neighbouring ridges (Vukusic et al., 1999). Such a high reflectance is remarkable if the presence of optically absorbing melanin, embedded in the scale material, is also taken into account. The advantage of the presence of melanin in this system is that it enhances the otherwise weaker contrast in RI between air and the cuticle material. Another feature of these type I scales in *Morpho* is the relative disorder in the structure, for example, in ridge heights, ridge alignment, lamellae periodicity, scale alignment and scale orientation. All this gives rise to an angularly more diffuse and broader reflectance (Kinoshita et al., 2002a,b). The reflection pattern from a single scale of *M. rhetenor* is divided into two separated wide-angle lobes. Rigorous optical examination of single scales showed that the broad angular spread in its reflection pattern is due to lateral tilting of the ridges and the lamellae leading to an asymmetry of the structure.

Whereas *M. rhetenor* exhibits a bright intense metallic blue appearance, *M. didius* has a more diffuse appearance. This butterfly species has two kinds of scales on the dorsal face of the wings (Fig. 11): ground scales, similar to the ones observed on *M. rhetenor*'s wings also with laterally tilted ridges, and

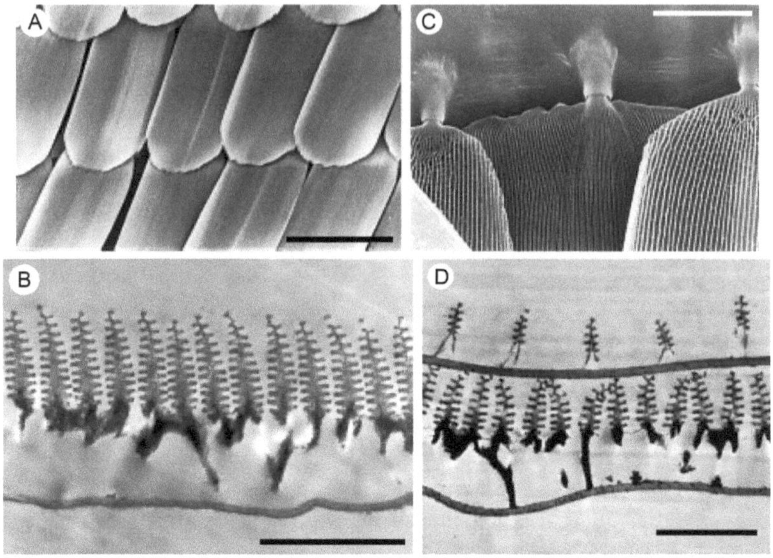

Fig. 11 Scales shingled on *M. rhetenor*'s wings and observed by SEM (A) and TEM (B). Ground and glass scales covering *M. didius'* iridescent wings imaged by SEM (C) and TEM (D). *Scale bars* correspond to 120 (A), 3 (B), 30 (C) and 2 μm (D). *Reproduced from Vukusic, P., Sambles, J.R., Lawrence, C.R., Wootton, R.J., 1999. Quantified interference and diffraction in single* Morpho *butterfly scales. Proc. R. Soc. B 266, 1403–1411.*

almost transparent covering scales, often referred to as *glass* scales, which completely overlap the ground scales (Ghiradella, 1994). Although the angular reflectance pattern from *M. didius*' scales differs from *M. rhetenor*, in the sense that there is no bi-lobed angular distribution, it is almost as spread. This difference in the observed single-scale reflectance distribution is achieved by the presence of glass scales. These strongly diffract light by transmission twice: first as light is incident upon the wing scales and second after it is subsequently reflected from the strongly reflecting blue scales underneath them. This broad diffraction-assisted angular scatter, as well as the fewer number of cuticle layers in the *M. didius* wing scale lamellae structure (up to eight layers) (Vukusic et al., 1999; Yoshioka and Kinoshita, 2004), leads to a less saturated colour with a peak reflectance of approximately 40%.

As mentioned previously, another important element in *Morpho*'s brilliant blue colour is the melanin content within the reflecting scale material. By absorbing wavelengths that are complementary to the bright blue reflected colours, this pigmentation reduces desaturating backscatter of nonblue colours, thereby increasing the colour saturation of its blue appearance (Vukusic et al., 1999; Yoshioka and Kinoshita, 2006a,b). Male *M. sulkowskyi* butterflies have a blue-silver colour appearance, whereas male *M. didius* butterflies exhibit a metallic blue appearance even though the photonic structures of their ground scales are very similar. This large difference in colour appearance is due to the presence of melanin absorbing green and red light and distributed in the lower parts of the ground scales of *M. didius* and the absence of such pigments in *M. sulkowskyi*'s scales (Kinoshita et al., 2002b). Interestingly, *M. sulkowskyi* was found to have some fluorophore pigmentation embedded within its scales. Its photonic nanostructures have been described as influencing the light emission from these fluorophores (Kumazawa et al., 1994; Van Hooijdonk et al., 2011).

In addition to these optical properties, characteristic type I nanostructuring of the scale ridges was demonstrated to give the scales hydrophobic and even superhydrophobic properties enabling a self-cleaning function (Berthier, 2010). Water, with its high surface tension, does not penetrate the structures: water droplets roll across and off the wing surfaces. This allows the butterfly wings to stay dry and reduce dust accumulation or contaminants. In contrast to water, however, ethanol and other alcohols have a low surface tension. Subsequently, when droplets of ethanol are deposited on *Morpho* wings, this liquid penetrates into the air gaps between ridge lamellae replacing the air in these gaps. This process reduced the inherent RI contrast of the scale system, thereby changing the conditions for optical interference and, on fluid penetration, inducing a green colour appearance (Vukusic et al., 1999; Wang et al., 2014). Similarly, the optical response of this photonic structure was found to be modified selectively upon exposure to very low concentrations of different vapours (Fig. 12) (Potyrailo et al., 2007). This selective colour change was explained by a differential adsorption onto the lamellae surfaces as a result of varying surface polarity gradient with ridge depth (Potyrailo et al., 2013).

4.2 Light Diffraction in *Lamprolenis nitida, Pierella luna* and *Argyrophorus argenteus*

Lamprolenis nitida is a butterfly species from the Papua New Guinea forests. The dorsal surfaces of both males and females exhibit a brownish appearance under diffuse broadband incident light. Nevertheless, in addition, a bright iridescent shine is observed on the male wings at a relatively grazing incidence (Ingram et al., 2008). With light incident from the anteroposterior

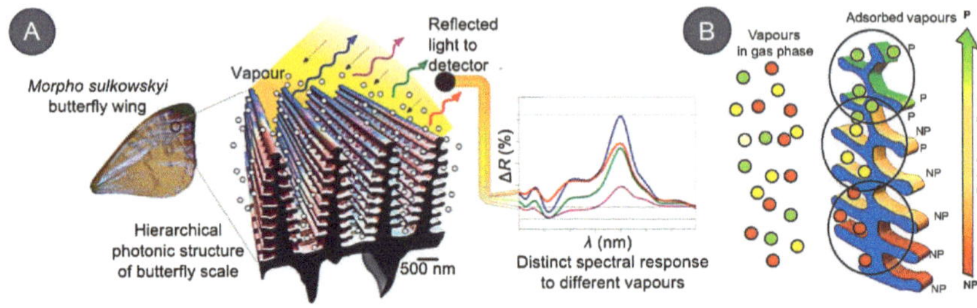

Fig. 12 (A) The optical response ΔR (change in reflectance) from *M. sulkowskyi* changes upon contact with different vapours and concentrations. (B) A gradient of surface polarity of the scale ridges runs from polar (P) tops down to the nonpolar (NP) bases, fostering preferential adsorption of vapours of different polarities onto the related regions of the ridges. *Panel (A): Reproduced from Potyrailo, R.A., Ghiradella, H., Vertiatchikh, A., Dovidenko, K., Cournoyer, J.R., Olson, E., 2007.* Morpho *butterfly wing scales demonstrate highly selective vapour response. Nat. Photonics 1, 123–128, with permission of Nature Publishing Group. Panel (B): Reproduced from Potyrailo, R.A., Naik, R.R., 2013. Bionanomaterials and bioinspired nanostructures for selective vapor sensing. Annu. Rev. Mater. Res. 43, 307–334.*

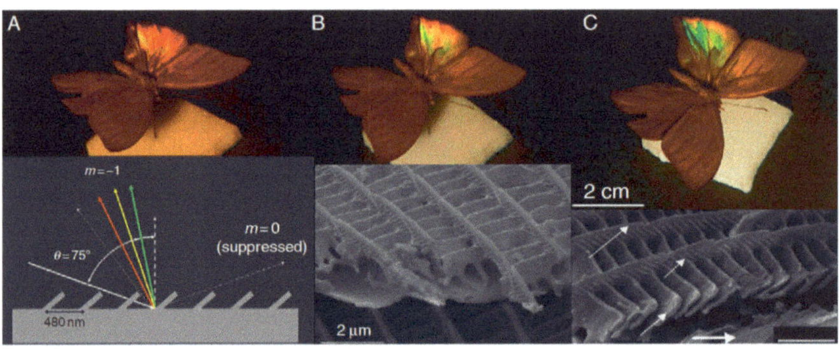

Fig. 13 The male *L. nitida* butterfly displays a bright iridescent shine ranging from *green* to *red* in an anteroposterior illumination (A–C) due to a blazed diffraction grating comprising the crossribs of the scales. The three *thin arrows* in the *low right figure* correspond to longitudinal ridges, fluted microribs and crossribs (from *top* to *bottom*). The *thick arrow* points towards the coastal margin. The scale bar corresponds to 2 μm. *Reproduced from Vigneron, J.-P., Simonis, P., 2010. Structural colours. Adv. Insect Physiol. 38, 181–218 with permission of Elsevier.*

direction, the hindwings exhibit an iridescent colour appearance that varies from green to red with increasing angle of observation (measured with respect to the normal to the wing surface) (Fig. 13). With a posteroanterior grazing illumination, a fainter iridescent colour is observed from blue to violet. The former optical behaviour was explained by considering the blazed diffraction grating formed by the periodic crossribs of its scales (Fig. 13).

These crossribs are tilted with an angle equal to 30° with respect to the scale surface and towards the butterfly body (Ingram et al., 2008). The latter optical response was found to be due to the periodic fluted microribs, tilted at 45° in the opposite direction, incorporated in the ridges and comprising another blazed grating (Fig. 13) (Ingram et al., 2008). The two iridescent colours were subsequently explained by two distinct blazed gratings on single scales. These blazed gratings were shown to be restricted to the diffraction order $m = -1$ (Eq. 7) and to suppress specular reflection ($m = 0$) (Ingram et al., 2008).

The forewings of the male *Pierella luna* butterfly exhibit a counterintuitive iridescence under posteroanterior illumination: varying from violet to red with a decreasing observation angle (Fig. 14), in contrast to the colour signature arising from a diffraction grating lying parallel to the wing membrane. In these scales, light diffraction arises from the crossribs. The reverse colour variation with the angle is due to curved scales covering the butterfly wings, leading in places to vertical gratings (Vigneron et al., 2010), i.e., perpendicular to the wing membrane (Fig. 14). These scales transmit incident light, instead of reflecting it. Since the incidence and detection angles are measured with respect to the grating axis, in this case, Eq. (7) becomes:

$$\lambda_{\text{dif}} = \frac{d(\cos\theta_{\text{inc}} + \cos\theta_{\text{dif}})}{m}. \tag{10}$$

Another example of a naturally evolved diffraction grating design lies in the scales of the wings of male *Argyrophorus argenteus* butterflies (Fig. 15), which exhibit a broadband silver colour appearance (Vukusic et al., 2009). Their scale windows are reduced. The scale nanostructures comprise variations in the periodicity of horizontally spaced small pores between the crossribs connecting neighbouring ridges (Vukusic et al., 2009). This quasi-disorder in periodicity leads to multicoloured strips along and between the scales' ridges that can be observed clearly by optical microscopy. Through additive colour mixing, the juxtaposition of these strips gives rise to the observed colour.

4.3 *Ancyluris meliboeus*

Type Ib scales exist on male *A. meliboeus* butterflies' wings. They exhibit an iridescent colour appearance that varies from blue to orange (Fig. 16) (Vukusic et al., 2001b, 2002). The lamellae are more tilted (30° with respect to the scale base) than the ones found on type Ia scales (Fig. 16A), leading to

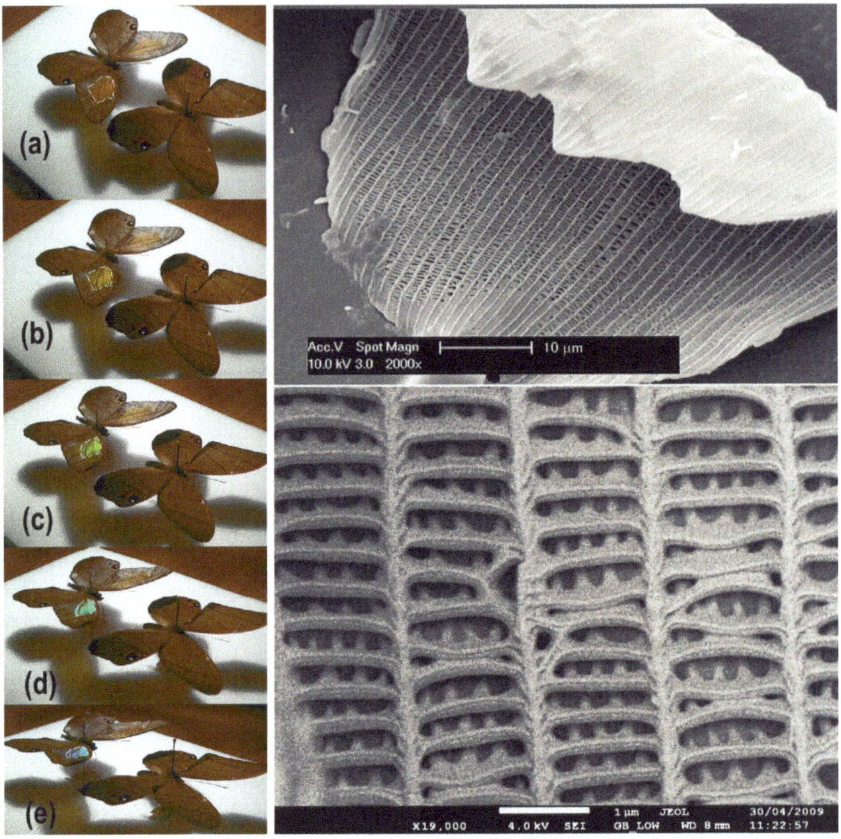

Fig. 14 Male (*left*) and female (*right*) *P. luna* butterflies (A–E) exhibit a *brown colour* appearance. The male displays an iridescent shine from *red* to *violet* with an increasing observation angle. This reverse colour order arises from the crossribs of curved scales lying on the wings and forming a vertical diffraction grating. *Reproduced from Vigneron, J.-P., Simonis, P., Aiello, A., Bay, A., Windsor, D.M., Colomer, J.-F., Rassart, M., 2010. Reverse color sequence in the diffraction of white light by the wing of the male butterfly* Pierella luna *(Nymphalidae: Satyrinae). Phys. Rev. E 82, 021903, with permission of the American Physical Society.*

the observed brilliant iridescence. These tilted lamellae give rise to two different optical responses that combine with each other. The ridge's upper surface comprises a grating which diffracts incident light in addition to the interference that arises within the lamellar multilayer observed within the ridges. Furthermore, due to the tilted feature of the nanostructure, no light whatsoever is scattered in a fraction of the observation hemisphere, a property that gives rise to a very fast and high contrast colour flicker signature (Fig. 16B) (Vukusic et al., 2001b, 2002).

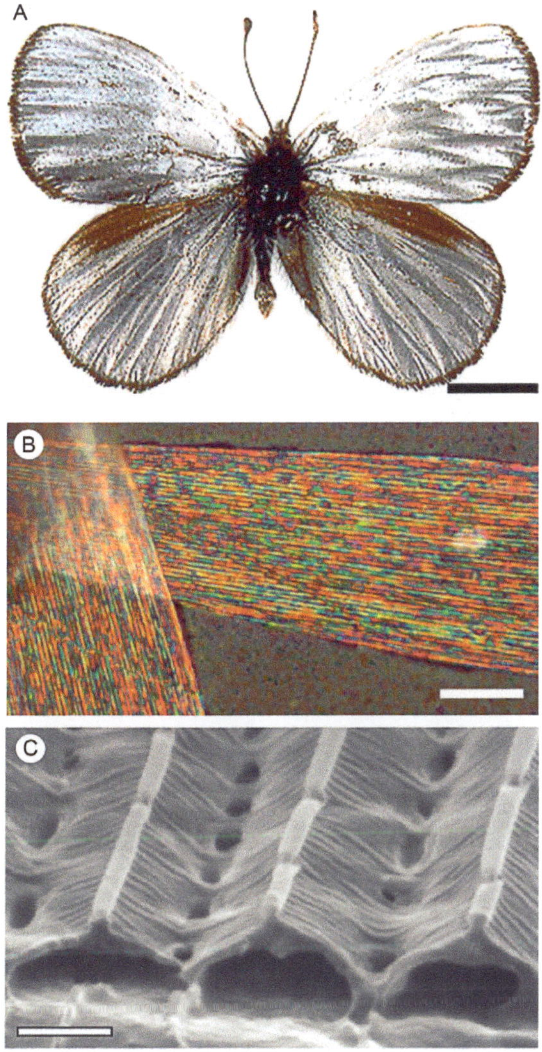

Fig. 15 *A. argenteus* butterfly wings display a *silver colour* (A) due to additive colour mixing from *colourful strips* between the scale ridges (B), arising from the quasi-disordered periodicity in the *reduced windows* and *crossribs* (C). The scale bars correspond to 0.5 cm (A), 15 μm (B) and 0.8 μm (C). *Reproduced from Vukusic, P., Kelly, R., Hooper, I., 2009. A biological sub-micron thickness optical broadband reflector characterized using both light and microwaves. J. R. Soc. Interface 6, S193–S201.*

4.4 *Troides magellanus*

The colour appearance of the hindwings of male *T. magellanus* butterflies comes from three sources in its type Ic scales, from pigments, from interference in scale ridge-based lamellae and from fluorescence (Lawrence et al.,

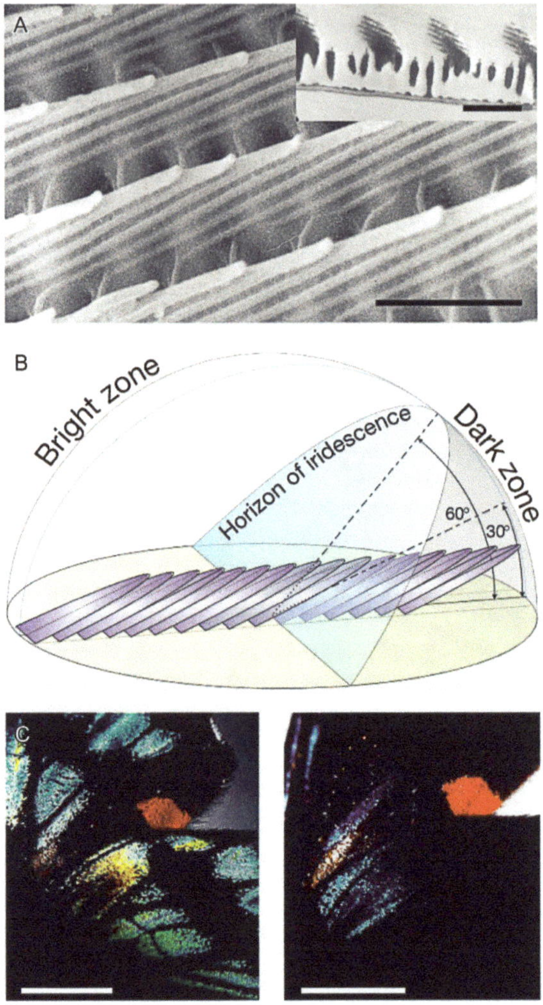

Fig. 16 The lamellae incorporated on *A. meliboeus* butterfly wing scales are tilted with respect to the basal lamina (A). Due to this inclination, a *dark zone* appears in the observation hemisphere (B). Within the *bright zone*, the wings are colourful and iridescent (C, *left*). These colours vanish within the *dark zone* (C, *right*). The *red area* is due to pigmentation. The scale bars correspond to 1 μm (A), 2 μm (*inset*) and 4 mm (C). *Reproduced from Vukusic, P., Sambles, J.R., Lawrence, C.R., Wootton, R.J., 2001. Structural colour: now you see it—now you don't. Nature 410, 36, with permission of Nature Publishing Group.*

2002; Van Hooijdonk et al., 2012c; Vigneron et al., 2008). To human observers, these wings have a striking uniform yellow colour appearance under broadband incident light (Fig. 17) due to the presence of yellow papiliochrome pigments in the scales covering both upperside and underside

Fig. 17 The hindwings of male *T. magellanus* butterfly exhibit a striking *yellow colouration* under white incident light. With a grazing incidence, a *bright blue-green* iridescent colour is observed. The lamellae on the ridges appear periodic ($b \cong 260$ nm) on SEM images and tilted by 60° with respect to the scale base. The *arrow* points towards the stalk of the scale and the insect body. The cross-section of ridges has a *triangular shape*, as observed by TEM. *Reproduced from Vigneron, J.-P., Kertész, K., Vértesy, Z., Rassart, M., Lousse, V., Bálint, Z., Biró, L.P., 2008. Correlated diffraction and fluorescence in the backscattering iridescence of the male butterfly* Troides magellanus *(Papilionidae). Phys. Rev. E 78, 021903, with permission of the American Physical Society.*

of the wing (Van Hooijdonk et al., 2012c). From the rear of the butterfly, at grazing incidence and observation, a bright blue-green iridescent reflection is observed (Fig. 17). Finally, under UV irradiation, papiliochrome II fluorescent pigments emit a yellow-green colouration. Electron micrographs show that the lamellae that form the ridges sit at an angle of 60° with respect to the scale substrate. Modelled as a bi-grating (Lawrence et al., 2002) or as a blazed diffraction grating (Vigneron et al., 2008), this structure was demonstrated in both cases to give rise to the observed blue-green iridescence (Fig. 17). The blazed grating approach also predicts an enhanced fluorescence emission from the fluorophores confined within the photonic structure (Van Hooijdonk et al., 2012a; Vigneron et al., 2008).

Interestingly, the forewing scales are very different from those on the hindwings. They are mainly black with a high absorption in the visible (up to 98%) and the infrared range due to the presence of melanin in their scales as well as light trapping within reported nanostructures on these scales (Herman et al., 2011).

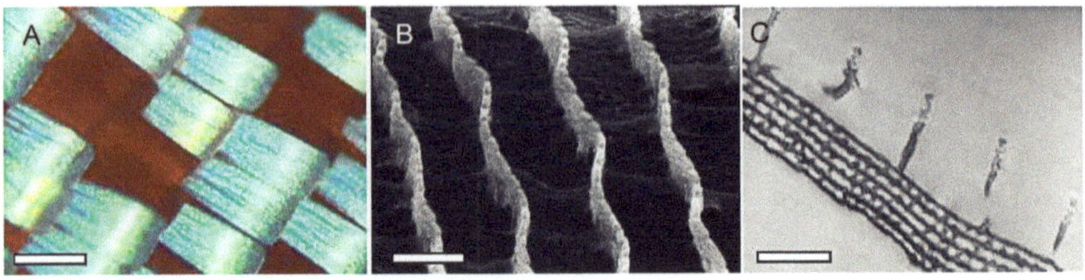

Fig. 18 *Greenish* and *bluish* iridescent scales cover *U. leilus* wings (A). The ridges (B) are believed to act as impedance-matching structure as well as diffraction grating. The multilayers (C) occurring in the scale bodies give rise to the observed structural colours. The scale bars correspond to 75 μm (A) and 1.2 μm (B, C). *Reproduced from Vukusic, P., Sambles, J.R., 2001. Shedding light on butterfly wings. Proc. SPIE 4438, 85–95.*

4.5 *Urania leilus* and *Chrysiridia rhipheus*

U. leilus and *Chrysiridia rhipheus* are two day-flying moth species from South America. The former exhibits greenish and bluish iridescent patches on both faces of its wings (Fig. 18). The latter is even more colourful and has wonderful patches ranging from blue to purplish red. The scales covering these wings belong to type IIa wing scale category, namely those giving rise to multilayer interference in the body of these scales (Lippert and Gentil, 1959; Yoshioka and Kinoshita, 2006a). The multilayers comprise a maximum of seven dense cuticle layers separated by porous largely air-filled structures (Fig. 18C). The crossribs are spaced with a period similar to the ridges spacing (Lippert and Gentil, 1959; Yoshioka and Kinoshita, 2007). The role of the ridges has been described as that of impedance-matching between the incidence medium and the layered scale region that reduces the broadband reflection from the scales as well as that of a diffraction grating to enhance broad-angle scattering from the multilayer (Vukusic, 2005; Vukusic et al., 2000a). It was found that the scale curvature in the longitudinal direction leads to multilength-scale colour mixing and polarisation effects due to light interaction between two neighbouring scales (Yoshioka and Kinoshita, 2007).

4.6 Multilayer Interference in Lycaenids' Scales

Many butterfly species from the Lycaenidae family have type IIa scales, the visual appearances of which are sometimes associated with both the ventral and dorsal sides of their wings. These wings are generally blue or green (Fig. 19A). Many lycaenids have been investigated: *Albulina metallica, Arhopala amantes, Arhopala japonica, Celastrina argiolus, Chrysozephyrus*

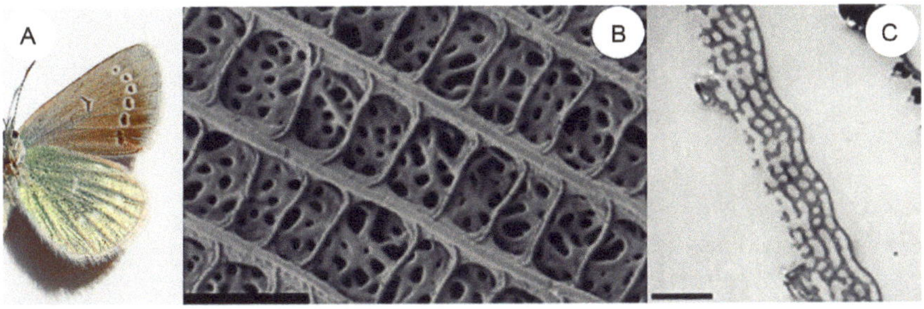

Fig. 19 The ventral side of the hindwings of *A. metallica* butterfly exhibit a *shiny green colour* (A) due to a perforated multilayer (B, C). The *brownish forewings* are pigmentary coloured. The scale bars in (B) and (C) correspond to 2 and 1 μm, respectively. *Panel (A): Reproduced from Kertész, K., Molnár, G., Vértesy, Z., Koós, A.A., Horváth, Z.E., Márk, G.I., Tapasztó, L., Bálint, Z., Tamáska, I., Deparis, O., Vigneron, J.-P., Biró, L.P., 2008. Photonic band gap materials in butterfly scales: a possible source of "blueprints". Mater. Sci. Eng. B 149, 259–265, with permission of Elsevier. Panels (B) and (C): Reproduced from Biró, L.P., Kertész, K., Vértesy, Z., Márk, G.I., Bálint, Z., Lousse, V., Vigneron, J.-P., 2007. Living photonic crystals: butterfly scales—nanostructure and optical properties. Mater. Sci. Eng. C 27, 941–946, with permission of Elsevier.*

aurorinus, *Chrysozephyrus brillantinus*, *Danis danis*, *Hypochrysops delicia*, *Jalmenus evagoras*, *Ogyris amaryllis*, *Plebejus icarioides*, *Polyommatus daphnis* and *Polyommatus icarus* (Biró et al., 2003, 2007; Kertész et al., 2008; Wilts et al., 2009). The multilayers found in their scales' bodies comprise alternating higher RI layers and lower RI layers (Fig. 19B and C). The former are often characterised by different degrees of so-called "pepper-pot" perforations and the latter are often made mostly of air with some spacer structures (Biró et al., 2003, 2007; Kertész et al., 2008; Wilts et al., 2009). Wilts and associates investigated 11 different lycaenid species and showed that the *perforation factor* ranges from 1% to 35% across the studied species (Wilts et al., 2009). These perforations result in a decrease of the effective RI of the higher RI layers, leading therefore to a smaller RI contrast between the two constituent layer types. This gives rise to the narrowing of the reflectance peaks and PBGs. The spatial scattering patterns of these butterflies vary but in general are diffuse. In spite of the fact that they seem to have not any obvious relationship with the perforation factor, these scattering patterns depend on the nanostructuring of the multilayers: presence of perforations, perforation sizes, perforation regularity and ridge geometry (Wilts et al., 2009). As is expected from the optical theory, the reflectance peak intensity and the peak wavelength were found to be functions of the layer number and the multilayer period, respectively. Interestingly, the presence of such multilayers in structurally coloured lycaenids was the result of environmental adaptation related to the thermoregulation of the butterflies' bodies (Biró et al., 2003).

4.7 Thin-Film Interference in *Graphium sarpedon* and *Hypolimnas salmacis*

Some of type IIa scales have a nanostructure in their body that is reduced to its minimum, i.e., an optical thin film. *Graphium sarpedon* is such an example (Stavenga, 2014; Stavenga et al., 2010, 2012) (Fig. 20A). On the ventral side of its wing, the so-called glass scales found in the bluish green patches contain no multilayer in their body but instead, the apposition of the scales' upper and lower laminae constitutes a 400-nm thin film (Fig. 20B and C). The scale internal structures and windows are completely missing. In addition to the presence of pigments in the underlying wing membrane, the thin film in the unpigmented glass scales' body leads to light interference (Stavenga, 2014; Stavenga et al., 2010, 2012). The small ridges and their connecting crossribs do not seem to play any role in the scales' appearance.

In addition to pigmentation embedded within their body that acts as a spectral filter, scales from nymphaline butterfly species such as *Aglais urticae*, *Aglais io* and *Vanessa atalanta* were found to give rise to such thin-film interference (Stavenga, 2014; Stavenga et al., 2014). In these cases, however, the upper and the lower laminae are separated by trabeculae. Only the 200-nm thick lower lamina gives rise to structural colour appearance. The stacking of neighbouring scales additionally increases the reflected light intensity. Ridges and crossribs found on the upper lamina are believed to act as light-diffusing elements.

Very recently, it was demonstrated that the iridescent blue patches on dorsal surfaces of the wings of *Hypolimnas salmacis* butterflies (Fig. 20D) are due to thin-film interference within the lower lamina of the type IIa covering glass scales stacked on absorbing brown scales (Fig. 20E–G) (Siddique et al., 2016). Variation in thickness across the span of the scale lower lamina leads to a weakening of the scale iridescence. Interestingly, a related species from the same genus, *Hypolimnas bolina*, the dorsal wing faces of which exhibit a similar iridescent blue, was reported to generate its colour appearance principally from type Ib scales (Kemp et al., 2014; Kemp and Macedonia, 2006).

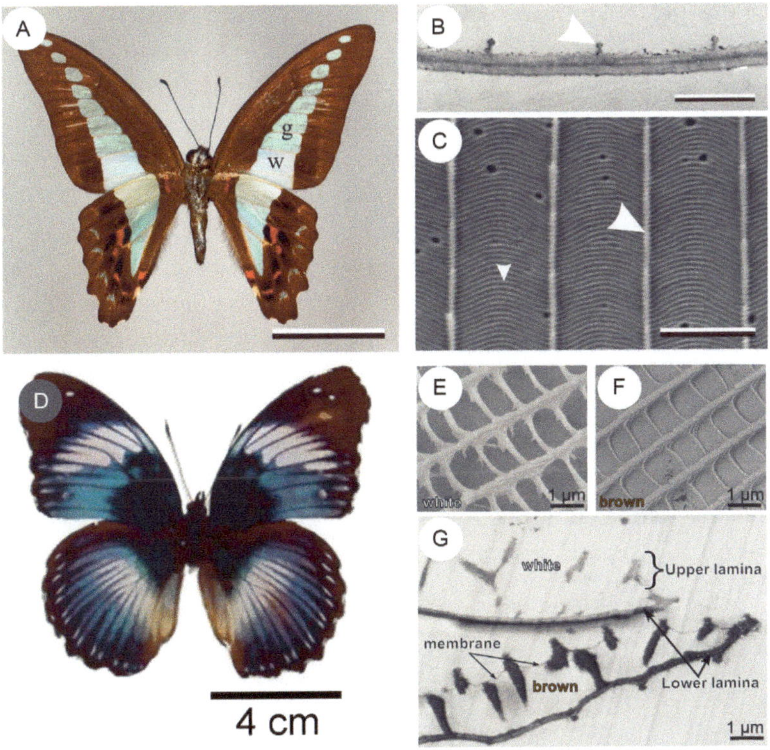

Fig. 20 The ventral faces of *G. sarpedon* butterfly exhibit *bluish green* (g) and *white* (w) *patches* (A). The glass scales covering the wing in *bluish green* areas consist of the apposition of the upper and lower laminae, forming a 400-nm thin film (B, C). The ridges (*large arrowheads*) and microribs (*small arrowhead*) do not play any specific role in the wing colour appearance. The *blue patches* on the dorsal face of *H. salmacis'* wings (D) are due to thin-film interference in the lower lamina of their glass scales (E, G) covering absorbing *brown scales* (F, G). The scale bars were unfortunately not mentioned in the original publication (A–C). *Panels (A)–(C): Reproduced from Stavenga, D.G., 2014. Thin film and multilayer optics cause structural colors of many insects and birds. Mater. Today Proc. 1S, 109–121, with permission of Elsevier. Panels (D)–(G): Reproduced from Siddique, R.H., Vignolini, S., Bartels, C., Wacker, I., Hölscher, H., 2016. Colour formation on the wings of the butterfly* Hypolimnas salmacis *by scale stacking. Sci. Rep. 6, 36204, under the version 4.0 of CC-BY licence.*

4.8 *Papilio ulysses* and *Papilio palinurus*

P. ulysses and *P. palinurus* butterflies, respectively, display blue and green colour patches of their wings due to the presence of type IIb scales. After the scales found on the species from the *Morpho* genus, these scales have been among the most investigated. They comprise approximately 10 air-separated cuticle layers. These air separations are supported by distributed small pillar-like structures (Fig. 21). The orthogonal modulations of these scale multilayers form distinct concavities between the ridges. The thickness of each layer remains almost constant across the entire scale span. The two main differences between *P. ulysses*' and *P. palinurus*' scales lie, first, in the period of the multilayer (i.e., smaller for the blue scales of *P. ulysses* and slightly larger for the green scales of *P. palinurus*) and, second, in the profiles of their scale concavities (i.e., deeper with a steeper gradient slope in the case of *P. palinurus* for which the concavity sides form an angle of about 45° with the base of the scale. The slopes of the concavities in the scales of *P. ulysses* are inclined by about 30° (Ghiradella, 1985; Vukusic et al., 2000b, 2001a).

Under broadband incident light, these differences in concavity profiles result in the iridescent reflection of yellow light in the centres of the concavities on *P. palinurus*' scales and in the flat areas between these concavities

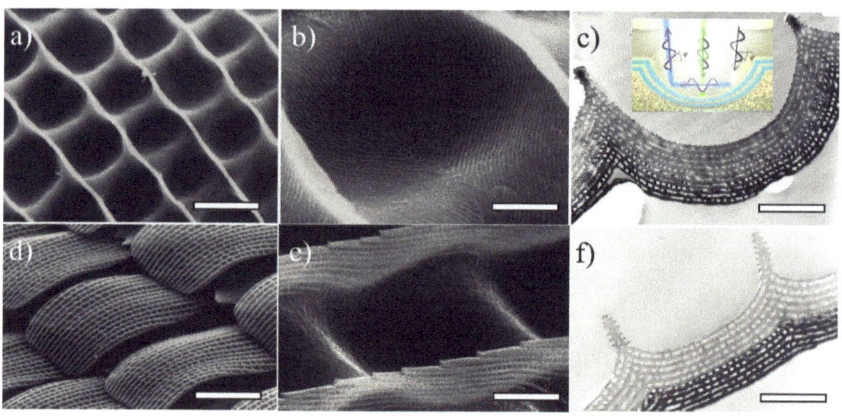

Fig. 21 SEM (A, B, D, E) and TEM (C, F) of the nanostructures found on *P. palinurus'* (A–C) and *P. ulysses'* (D–F) scales. The multilayers incorporated within the scale bodies give rise to structural colour appearances. Due to the shape of the concavity of *P. palinurus'* scales, linear polarisation conversion of incident *blue light* occurs on the concavity sides (C, *inset*). The scale bars correspond to 5 μm (A), 1 μm (B, C, E, F) and 6 μm (D). *Panels (A)–(F): Reproduced from Vukusic, P., Sambles, J.R., 2001. Shedding light on butterfly wings. Proc. SPIE 4438, 85–95. Panel (C, inset): Reproduced from Kolle, M., Salgard-Cunha, P.M., Scherer, M.R.J., Huang, F., Vukusic, P., Mahajan, S., Baumberg, J.J., Steiner, U., 2010. Mimicking the colourful wing scale structure of the* Papilio blumei *butterfly. Nat. Nanotechnol. 5, 511–515.*

and blue light reflection on the side walls. In contrast to this, only iridescent blue light is observed at the centres of *P. ulysses*' concavities under normally incident broadband light with the concavity sides remaining dark. Under a diffuse white incident light and with a grazing observation, *P. palinurus*' wings appear blue (Huxley, 1975; Vukusic et al., 2000b). At normal incidence, *P. palinurus*' and *P. ulysses*' multilayers reflect yellow and blue colours, respectively, due to the difference in period. Such a reflection occurs at the concavity centres. Because of the 45° inclination of the concavity sides, normally incident light interacting with such a side in *P. palinurus*' scales is reflected towards the opposite side of the same concavity, also inclined by 45°, where it is eventually reflected back in the direction from which it originated (Fig. 21) (Vukusic et al., 2000b, 2001a). Since *P. ulysses*' concavity sides are not orthogonal, this reflection property does not occur. Furthermore, with respect to the *P. palinurus*' concavity structure, this double reflection gives rise to linear polarisation conversion of incident blue light (Fig. 21C, inset) and a possible role in linear polarisation signalling (Vukusic et al., 2000b, 2001a). With an increasing observation angle, the double reflection from two opposite sides of the concavities decreases since the angle-dependent reflectance spectra of each side are mismatched. To human vision the combination of yellow and blue reflections from *P. palinurus*' scales and wings gives rise to the species' green wing appearance through additive colour mixing: this leads to the possibility of a crypsis function in addition to the polarisation signalling (Vukusic et al., 2000b, 2001a). The blue colour of *P. ulysses* is believed to foster long-range visibility. Its iridescence is diminished by the concavities (Prum et al., 2006; Vukusic et al., 2001a). The ridges do not play a significant role in these butterflies' colour appearance although an impedance-matching function is believed to arise from these structural elements (Vukusic, 2005). Similar to the case of *T. magellanus*, black regions on *P. ulysses* wings were found to be highly absorptive due to the nanostructure of these scales (Vukusic et al., 2004).

The visual appearance of *P. palinurus* and similar visual effects found in related butterfly such as *Papilio blumei* are difficult to mimic synthetically. In addition, a recent study showed that *P. blumei*'s wings exhibit a more striking colour change (with respect to both intensity variations and reflectance peak wavelength shift) upon contact with ethanol than *M. rhetenor*'s wings (Wang et al., 2014). That is why many research groups have tried to reproduce such structures in order to develop applications such as anti-counterfeiting of banknotes or sensing through a bioinspiration approach (Berthier et al., 2007; Gaillot et al., 2008; Kolle et al., 2010; Rasson et al., 2017).

4.9 Coherent Light Scattering in 3D PCs in *Callophrys rubi* and *Parides sesostris*

The type IIIa scales found on *C. rubi* (Allyn and Downey, 1976; Michielsen and Stavenga, 2008; Michielsen et al., 2010; Morris, 1975), *P. sesostris* (Michielsen and Stavenga, 2008; Prum et al., 2006; Saranathan et al., 2010; Vukusic and Sambles, 2001) as well as other butterflies from the Papilionidae and Lycaenidae families such as *Teinopalpus imperialis* (Argyros et al., 2002; Michielsen and Stavenga, 2008; Saranathan et al., 2010), *Cyanophrys remus* (Kertész et al., 2006; Michielsen and Stavenga, 2008), *Cyanophrys herodotus* (Saranathan et al., 2010), *Callophrys gryneus* (formerly, *Mitoura gryneus*) (Michielsen and Stavenga, 2008; Prum et al., 2006; Saranathan et al., 2010) and *Callophrys dumetorum* (Michielsen and Stavenga, 2008; Prum et al., 2006; Saranathan et al., 2010) are known to comprise ordered 3D PCs between the upper and lower laminae of their wing scales (Fig. 22). These structures scatter incident light with specific wavelengths into certain directions leading to a range of structural colour appearances. They are most often diffuse and green in appearance.

In 2008, following the suggestion of several different photonic models for the prediction of these structures' optical responses, it was demonstrated that these PCs could be very accurately modelled by single gyroid structures characterised by different lattice parameters and filling fractions (i.e., the ratio between the volume occupied by cuticle material and the total volume) (Michielsen and Stavenga, 2008; Saranathan et al., 2010). Gyroid structures comprise two different regions separated by a so-called bicontinuous triply periodic minimal surface (Fig. 23). This surface minimises its surface area under some constrains. The *bicontinuity* implies that these two regions are continuous. They are, however, not simply connected but interpenetrate each other in a complex way. Mathematically in 3D space, a gyroid minimal surface may be defined with the function:

$$F(X, Y, Z) = \sin X \cos Y + \sin Y \cos Z + \cos X \sin Z = t, \qquad (11)$$

where t is a parameter so that the filling factor is a function of it; $X = 2\pi x/a$, $Y = 2\pi y/a$ and $Z = 2\pi z/a$ with the space coordinates x, y, z and the lattice parameter is a. The RI function $n(x, y, z)$ is usually chosen so that $n(X, Y, Z)$

Fig. 22 Ventral wing scales of *C. remus* butterfly. PC domains are incorporated in the region between the upper ridge lamina and the basal lamina (A, B). The photonic polycrystal gives rise to an angle-independent structural colour due to additive colour mixing. *Reproduced from Vigneron, J.-P., Simonis, P., 2010. Structural colours. Adv. Insect Physiol. 38, 181–218, with permission of Elsevier.*

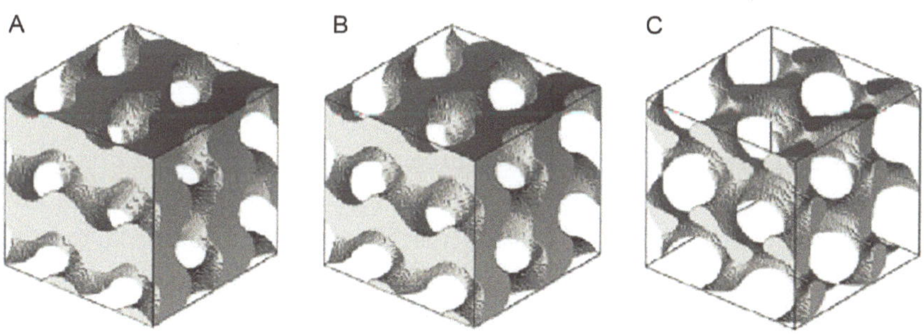

Fig. 23 Eight unit cells of gyroid structures with parameter t (Eq. 11) equal to 0.0 (A), -0.3 (B) and -1.0 (C) corresponding to filling factors equal to 50%, 40% and 17%, respectively. *Reproduced from Michielsen, K., Stavenga, D.G., 2008. Gyroid cuticular structures in butterfly wing scales: biological photonic crystals. J. R. Soc. Interface 5, 85–94, under the version 4.0 of CC-BY licence.*

is equal to the RI of the cuticle material if $F(X, Y, Z) < t$ and is equal to 1 (air) if $F(X, Y, Z) \geq t$. The two bicontinuous regions are therefore filled either with cuticle material or with air: in the case of the naturally occurring butterfly gyroid, the air region is the one that occupies the largest fraction of gyroid's volume. The Bravais lattice symmetry of the resulting structure is body-centred cubic. The lepidopteran species investigated so far have been found with filling fractions ranging from 17% (*C. rubi*) to 40% (*P. sesostris*). With a parameter $t = -1$, *C. rubi*'s photonic structure appears optimised for creating the largest PBG extension in gyroid structures in butterfly wings.

The formation of the associated complex wing scale membrane was suggested as resulting from the interaction of the cell plasma membrane and the smooth endoplasmic reticulum membrane in each scale-forming cell (Ghiradella, 1998; Michielsen and Stavenga, 2008; Saranathan et al., 2010).

Within the scale, such PCs are usually separated into domains, the size of which is typically on the order of a few microns. The close juxtaposition of domains with different orientations gives rise to noniridescent and diffusely scattered colours that form as a result of additive colour mixing (Kertész et al., 2006; Morris, 1975). Noniridescent diffuse green visual appearances like these are believed to play a role in crypsis.

4.10 Incoherent Light Scattering in *Pieris rapae* and *Delias nigrina*

Ovoid bead structures found in the open spaces between the ridges and crossribs of the type IIIb scales of *P. rapae* and *D. nigrina* butterflies were found to give rise to the matt white appearance of these butterflies (Stavenga et al., 2004, 2006). These random ovoid-shaped scattering elements (Fig. 24), the sizes of which are of the order of a few hundred of nanometres, diffusely scatter incident broadband light in all directions, leading to a white colour appearance that is independent of the direction of observation. The reflectance intensity was found to be a function of the bead density (Luke et al., 2009; Morehouse et al., 2007; Stavenga et al., 2004). The appearance from these two butterflies' wings was shown to be brighter than from the white patches of the wings of *Heliconius melpomene*, the scales of which do not comprise such analogous scattering pigment beads (Stavenga et al., 2004).

Such scattering structures can be used in combination with pigmentation such as in the case of male *P. rapae* butterflies where leucopterin and xanthopterin pigments absorb incident ultraviolet light (Makino et al., 1952; Yagi, 1954). Absorbing pigments embedded in the scales of these

Fig. 24 Randomly disordered scatterers (granules) in the scale bodies of male *Pontia protodice* dorsal wings. The scale bar corresponds to 1 μm. *Reproduced from Morehouse, N.I., Vukusic P., Rutowski, R., 2007. Pterin pigment granules are responsible for both broadband light scattering and wavelength selective absorption in the wing scales of pierid butterflies. Proc. R. Soc. B 274, 359–366.*

butterflies' wings strongly filter light in a manner that depends on pigment density. Such pigmentation was also identified in the white sections of the dorsal faces of *H. melpomene*'s wings (Stavenga et al., 2004). Interestingly, the female *P. rapae crucivora* butterfly's scales were found to lack these pigments and to exhibit a high reflectance in the UV range (Obara, 1970). Because of the butterflies' spectral sensitivity to UV, such differences can play a strong role in conspecific signalling and visual recognition: specimens from this species may see males with a specific colour that is very different to that of the females. In addition to the white scales present on its dorsal wing faces, yellow and red stripes are found on the ventral wing faces of *D. nigrina*. The scales found in these stripes also exhibit ovoid beads. However, these are combined with blue absorbing pigments as well as blue and yellow absorbing pigments, respectively. The black patches on *P. rapae*'s wings, the black tips of *D. nigrina*'s dorsal wing faces and the brown sections of *D. nigrina*'s ventral wing faces are due to the broadband optical absorption of melanin pigments.

In addition to this pigmentary absorption of UV, light reflectance was found to be increased by multiple scattering between cover and ground scales from both faces of the wings of *P. rapae* (Stavenga et al., 2006). Such multiscattering in stacked scales comprising randomly distributed bead structures and UV absorbing pigmentation was also demonstrated to be responsible for the colour appearance of the bright white dorsal faces of *D. nigrina* (Stavenga et al., 2006). The red scales found in the related stripes on the ventral face of its wings have a nonnegligible contribution to the dorsal reflectance in the red range of the electromagnetic spectrum (Stavenga et al., 2006).

5. CONCLUSION

Photonic structures found in lepidopteran scales are particularly sophisticated optical systems that give rise to striking visual effects, among them brilliant and varied structural colours. These colour appearances are due to coherent and incoherent light scattering within structures present in the form of single thin films, multilayers, diffraction gratings, 2D and 3D PCs and disordered structures mainly comprising air and cuticle. A broad range of optical effects and signatures, which cannot be elicited only using pigmentation, is achieved by these structures. These comprise iridescence, narrow-band reflection, large solid-angle scattering, polarisation effects, additive colour mixing and more. They are responsible for a range of biological functions: conspicuous appearances such as bright colours and scattering in large solid angles are usually related to conspecific recognition and courtship, whereas inconspicuous colours, e.g., low reflection surface and angle-independent colours, are often for crypsis.

In this chapter, lepidopteran scales' classification into three main categories and subcategories were outlined. These classes were described on the basis of certain morphological elements in the scales that are responsible for the related structural colour appearances. Several examples from these categories were reviewed, allowing a direct comparison between scales from different species with the intention of presenting an overview of the fundamental optical and photonic processes at the heart of the exhibited colours.

ACKNOWLEDGEMENTS

S.R.M. was supported by Wallonia-Brussels International (WBI) through a Postdoctoral Fellowship for Excellence program WBI·WORLD.

REFERENCES

Allyn, A.C., Downey, J.C., 1976. Diffraction structures in the wing scales of *Callophrys (Mitoura) siva siva* (Lycaenidae). Bull. Allyn Mus. 40, 1–6.

Argyros, A., Manos, S., Large, M.C.J., McKenzie, D.R., Cox, G.C., Dwate, D.M., 2002. Electron tomography and computer visualisation of a three-dimensional 'photonic' crystal in a butterfly wing-scale. Micron 33, 483–487.

Arnold, K.E., Owens, I.P.F., Marshall, N.J., 2002. Fluorescent signaling in parrots. Science 295, 92.
Berthier, S., 2003. Iridescences, les couleurs physiques des insectes, first ed. Springer, Paris.
Berthier, S., 2010. Photonique des *Morphos*, first ed. Springer, Paris.
Berthier, S., Charron, E., Boulenguez, J., 2006. Morphological structure and optical properties of the wings of Morphidae. Insect Sci. 13, 145–157.
Berthier, S., Boulenguez, J., Bálint, Z., 2007. Multiscaled polarization effects in *Suneve coronata* (Lepidoptera) and other insects: application to anti-counterfeiting of banknotes. Appl. Phys. A 86, 123–130.
Birch, K.P., Downs, M.J., 1993. An updated Edlén equation for the refractive index of air. Metrologia 30, 155–162.
Birch, K.P., Downs, M.J., 1994. Correction to the updated Edlén equation for the refractive index of air. Metrologia 31, 315–316.
Biró, L.P., Vigneron, J.-P., 2011. Photonic nanoarchitectures in butterflies and beetles: valuable sources for biosinspiration. Laser Photonics Rev. 5, 27–51.
Biró, L.P., Bálint, Z., Kertész, K., Vértesy, Z., Márk, G.I., Horváth, Z.E., Balázs, J., Méhn, D., Kiricsi, I., Lousse, V., Vigneron, J.-P., 2003. Role of photonic-crystal-type structures in the thermal regulation of a Lycaenid butterfly sister species pair. Phys. Rev. E Stat. Nonlin. Soft Matter Phys. 67, 021907.
Biró, L.P., Kertész, K., Vértesy, Z., Márk, G.I., Bálint, Z., Lousse, V., Vigneron, J.-P., 2007. Living photonic crystals: butterfly scales—nanostructure and optical properties. Mater. Sci. Eng. C 27, 941–946.
Biró, L.P., Kertész, K., Vértesy, Z., Bálint, Z., 2008. Photonic nanoarchitectures occurring in butterfly scales as selective gas/vapor sensors. Proc. SPIE 7057, 705706.
Born, M., Wolf, E., 1975. Principles of Optics, fifth ed. Pergamon Press, Oxford.
Boroditsky, M., Krauss, T.F., Coccioli, R., Vrijen, R., Bhat, R., Yablonovitch, E., 1999. Light extraction from optically pumped light-emitting diode by thin-slab photonic crystals. Appl. Phys. Lett. 75, 1036–1038.
Boulenguez, J., Berthier, S., Leroy, F., 2012. Multiple scaled disorder in the photonic structure of *Morpho rhetenor* butterfly. Appl. Phys. A 106, 1005–1011.
Burresi, M., Cortese, L., Pattelli, L., Kolle, M., Vukusic, P., Wiersma, D.S., Steiner, U., Vignolini, S., 2014. Bright-white beetle scales optimise multiple scattering of light. Sci. Rep. 4, 6075.
Catala-Stucki, R., 1959. Fluorescence effects from corals irradiated with ultra-violet rays. Nature 183, 949.
Chamberlin, G.J., Chamberlin, D.G., 1980. Colour: Its Measurement, Computation and Application. Heyden International Topics in Science, London.
Chen, J.I.L., von Freymann, G., Choi, S.Y., Kitaev, V., Ozin, G.A., 2006. Amplified photochemistry with slow photons. Adv. Mater. 18, 1915–1919.
Chen, J.I.L., von Freymann, G., Choi, S.Y., Kitaev, V., Ozin, G.A., 2008. Slow photons in the fast lane in chemistry. J. Mater. Chem. 18, 369–373.
Cockayne, E.A., 1924. The distribution of fluorescent pigments in Lepidoptera. T. R. Ent. Soc. Lond. 72, 1–19.
Colomer, J.-F., Simonis, P., Bay, A., Cloetens, P., Suhonen, H., Rassart, M., Vandenbem, C., Vigneron, J.-P., 2012. Photonic polycrystal in the greenish-white scales of the African longhorn beetle *Prosopocera lactator* (Cerambycidae). Phys. Rev. E Stat. Nonlin. Soft Matter Phys. 85, 011907.
Davies, A.J., 1980. The Finite Element Method: A First Approach, first ed. Clarendon Press, Oxford.
Demésy, G., John, S., 2012. Solar energy trapping with modulated silicon nanowire photonic crystals. J. Appl. Phys. 112, 074326.
Denton, E., 1971. Reflectors in fishes. Sci. Am. 224, 64–72.

Deparis, O., Vigneron, J.-P., 2010. Modeling the photonic response of biological nanostructures using the concept of stratified medium: the case of a natural three-dimensional photonic crystal. Mater. Sci. Eng. B 169, 12–15.
Deparis, O., Vandenbem, C., Rassart, M., Welch, V.L., Vigneron, J.-P., 2006. Color-selecting reflectors inspired from biological periodic multilayer structures. Opt. Express 14, 3547–3555.
Deparis, O., Khuzayim, N., Parker, A., Vigneron, J.-P., 2009. Assessment of the anti-reflection property of moth wings by three-dimensional transfer-matrix optical simulations. Phys. Rev. E Stat. Nonlin. Soft Matter Phys. 79, 041910.
Deparis, O., Mouchet, S.R., Dellieu, L., Colomer, J.-F., Sarrazin, M., 2014. Nanostructured surfaces: bioinspiration for transparency, coloration and wettability. Mater. Today Proc. 1S, 122–129.
Deparis, O., Mouchet, S.R., Su, B.-L., 2015. Light harvesting in photonic crystals revisited: why do slow photons at the blue edge enhance absorption? Phys. Chem. Chem. Phys. 17, 30525–30532.
Dhatt, G., Touzot, G., 1984. The Finite Element Method Displayed, first ed. John Wiley & Sons, Chichester.
Edlén, B., 1966. The refractive index of air. Metrologia 2, 71–80.
Erchak, A.A., Ripin, D.J., Fan, S., Rakich, P., Joannopoulos, J.D., Ippen, E.P., Petrich, G.S., Kolodziejski, L.A., 2001. Enhanced coupling to vertical radiation using a two-dimensional photonic crystal in a semiconductor light-emitting diode. Appl. Phys. Lett. 78, 563–565.
Foresi, J.S., Villeneuve, P.R., Ferrera, J., Thoen, E.R., Steinmeyer, G., Fan, S., Joannopoulos, J., Kimerling, L.C., Smith, H.I., Ippen, E.P., 1997. Photonic-bandgap microcavities in optical waveguides. Nature 390, 143–145.
Gaillot, D.P., Deparis, O., Welch, V., Wagner, B.K., Vigneron, J.-P., Summers, C.J., 2008. Composite organic-inorganic butterfly scales: production of photonic structures with atomic layer deposition. Phys. Rev. E Stat. Nonlin. Soft Matter Phys. 78, 031922.
Galusha, J.W., Richey, L.R., Gardner, J.S., Cha, J.N., Bartl, M.H., 2008. Discovery of a diamond-based photonic crystal structure in beetle scales. Phys. Rev. E Stat. Nonlin. Soft Matter Phys. 77, 050904.
Gao, Y., Xia, Q., Liao, G., Shi, T., 2011. Sensitivity analysis of a bioinspired refractive index based gas sensor. J. Bionic Eng. 8, 323–334.
Ghiradella, H., 1985. Structure and development of iridescent lepidopteran scales: the Papilionidae as a showcase family. Ann. Entomol. Soc. Am. 78, 252–264.
Ghiradella, H., 1989. Structure and development of iridescent butterfly scales: lattices and laminae. J. Morphol. 202, 69–88.
Ghiradella, H., 1994. Structure of butterfly scales: patterning in an insect cuticle. Microsc. Res. Tech. 27, 429–438.
Ghiradella, H., 1998. Hairs, bristles and scales. In: Harrison, F.W., Locke, M. (Eds.), Microscopic Anatomy of Invertebrates. In: Insecta, vol. 11A. John Wiley & Sons, New York, pp. 257–287.
Goodwin, R.H., 1953. Fluorescent substances in plants. Annu. Rev. Plant Physiol. 4, 283–304.
Hecht, E., 1998. Optics, third ed. Addison Wesley Longman, Reading.
Herman, A., Vandenbem, C., Deparis, O., Simonis, P., Vigneron, J.-P., 2011. Nanoarchitecture in the black wings of *Troides magellanus*: a natural case of absorption enhancement in photonic materials. Proc. SPIE 8094, 80940H.
Holt, A.L., Vahidinia, S., Gagnon, Y.L., Morse, D.E., Sweeney, A.M., 2014. Photosymbiotic giant clams are transformer of solar flux. J. R. Soc. Interface 11, 20140678.
Huxley, A.F., 1968. A theoretical treatment of the reflexion of light by multilayer structures. J. Exp. Biol. 48, 227–245.

Huxley, J., 1975. The basis of structural colour variation in two species of *Papilio*. J. Ent. Ser. A 50, 9–22.
Huxley, J., 1976. The coloration of *Papilio zalmoxis* and *P. antimachus*, and the discovery of Tyndall blue in butterflies. Proc. R. Soc. B 193, 441–453.
Ingram, A.L., Lousse, V., Parker, A.R., Vigneron, J.-P., 2008. Dual gratings interspersed on a single butterfly scale. J. R. Soc. Interface 5, 1387–1390.
Ingram, A.L., Ball, A.D., Parker, A.R., Deparis, O., Boulenguez, J., Berthier, S., 2009. Characterization of the green iridescence on the chelicerae of the tube web spider, *Segestria florentina* (Rossi 1790) (Araneae, Segestriidae). J. Arachnol. 37, 68–71.
Ingram, A.L., Deparis, O., Boulenguez, J., Kennaway, G., Berthier, S., Parker, A.R., 2011. Structural origin of the green iridescence on the chelicerae of the red-backed jumping spider, *Phidippus johnsoni* (Salticidae: Araneae). Arthropod Struct. Dev. 40, 21–25.
Israelowitz, M., Rizvi, S.H.W., von Schroeder, H.P., 2007. Fluorescence of the "fire-chaser" beetle, *Melanophila acuminate*. J. Lumin. 126, 149–154.
Jiang, T., Peng, Z., Wu, W., Shi, T., Liao, G., 2014. Gas sensing using hierarchical micro/nanostructures of *Morpho* butterfly scales. Sens. Actuators A Phys. 213, 63–69.
Joannopoulos, J.D., Johnson, S.G., Winn, J.N., Meade, R.D., 2008. Photonic Crystals: Molding the Flow of Light, second ed. Princeton University Press, Princeton.
John, S., 1987. Strong localization of photons in certain disordered dielectric superlattices. Phys. Rev. Lett. 58, 2486–2489.
Johnson, S.G., Joannopoulos, J.D., 2001. Block-iterative frequency-domain methods for Maxwell's equations in a planewave basis. Opt. Express 8, 173–190.
Judd, D.B., Wyszecki, G., 1975. Color in Business, Science and Industry. John Wiley & Sons, New York.
Kemp, D.J., Macedonia, J.M., 2006. Structural ultraviolet ornamentation in the butterfly *Hypolimnas bolina* L. (Nymphalidae): visual, morphological and ecological properties. Aust. J. Zool. 54, 235–244.
Kemp, D.J., Jones, D., Macedonia, J.M., Krockenberger, A.K., 2014. Female mating preferences and male signal variation in iridescent *Hypolimnas* butterflies. Anim. Behav. 87, 221–229.
Kertész, K., Bálint, Z., Vértesy, Z., Márk, G.I., Lousse, V., Vigneron, J.-P., Rassart, M., Biró, L.P., 2006. Gleaming and dull surface textures from photonic-crystal-type nanostructures in the butterfly *Cyanophrys remus*. Phys. Rev. E Stat. Nonlin. Soft. Matter Phys. 74, 021922.
Kertész, K., Molnár, G., Vértesy, Z., Koós, A.A., Horváth, Z.E., Márk, G.I., Tapasztó, L., Bálint, Z., Tamáska, I., Deparis, O., Vigneron, J.-P., Biró, L.P., 2008. Photonic band gap materials in butterfly scales: a possible source of "blueprints" Mater. Sci. Eng. B 149, 259–265.
Kertész, K., Piszter, G., Jakab, E., Bálint, Z., Vértesy, Z., Biró, L.P., 2013. Color change of blue butterfly wing scales in an air—vapor ambient. Appl. Surf. Sci. 281, 49–53.
Kertész, K., Piszter, G., Jakab, E., Bálint, Z., Vértesy, Z., Biró, L.P., 2014. Temperature and saturation dependence in the vapor sensing of butterfly wing scales. Mater. Sci. Eng. C 39, 221–226.
Kinoshita, S., 2008. Structural Colors in the Realm of Nature, first ed. World Scientific Publishing Co, Singapore.
Kinoshita, S., Yoshioka, S., Fujii, Y., Okamoto, N., 2002a. Photophysics of structural color in the *Morpho* butterflies. Forma 17, 103–121.
Kinoshita, S., Yoshioka, S., Kawagoe, K., 2002b. Mechanisms of structural colour in the *Morpho* butterfly: cooperation of regularity and irregularity in an iridescent scale. Proc. R. Soc. B 269, 1417–1421.
Kolle, M., Salgard-Cunha, P.M., Scherer, M.R.J., Huang, F., Vukusic, P., Mahajan, S., Baumberg, J.J., Steiner, U., 2010. Mimicking the colourful wing scale structure of the *Papilio blumei* butterfly. Nat. Nanotechnol. 5, 511–515.

Kumazawa, K., Tanaka, S., Negita, K., Tabata, H., 1994. Fluorescence from wing of *Morpho sulkowskyi* butterfly. Jpn. J. Appl. Phys. 33, 2119–2122.
Lafait, J., Andraud, C., Berthier, S., Boulenguez, J., Callet, P., Dumazet, S., Rassart, M., Vigneron, J.-P., 2010. Modeling the vivid white color of the beetle *Calothyrza margaritifera*. Mater. Sci. Eng. B 169, 16–22.
Lagorio, M.G., Cordon, G.B., Iriel, A., 2015. Reviewing the relevance of fluorescence in biological systems. Photochem. Photobiol. Sci. 14, 1538–1559.
Land, M.F., 1972. The physics and biology of animal reflectors. Prog. Biophys. Mol. Biol. 24, 75–106.
Lawrence, R.F., 1954. Fluorescence in arthropoda. J. Ent. Soc. South Africa 17, 167–170.
Lawrence, C., Vukusic, P., Sambles, R., 2002. Grazing-incidence iridescence from a butterfly wing. Appl. Opt. 41, 437–441.
Lee, D.W., Lowry, J.B., 1975. Physical basis and ecological significance of iridescence in blue plants. Nature 254, 50–51.
Lippert, W., Gentil, K., 1959. Über lamellare Feinstrukturen bei den Schillerschuppen der Schmetterlinge vom *Urania*- und *Morpho*-Typ. Z. Morphol. Ökol. Tiere 48, 115–122.
Liu, F., Dong, B.Q., Liu, X.H., Zheng, Y.M., Zi, J., 2009. Structural color change in longhorn beetles *Tmesisternus isabellae*. Opt. Express 17, 16183–16191.
Luke, S.M., Vukusic, P., Hallam, B., 2009. Measuring and modelling optical scattering and the colour quality of white pierid butterfly scales. Opt. Express 17, 14729–14743.
Makino, K., Satoh, K., Koiki, M., Ueno, N., 1952. Sex in *Pieris rapae* L. and the pteridin content of their wings. Nature 170, 933–934.
Marshall, J., Johnsen, S., 2017. Fluorescence as a means of colour signal enhancement. Philos. Trans. R Soc. Lond B Biol. Sci. 372, 20160335.
Mason, C.W., 1926. Structural colors in insects. I. J. Phys. Chem. 30, 383–395.
Mäthger, L.M., Land, M.F., Siebeck, U.E., Marshall, N.J., 2003. Rapid colour changes in multilayer reflecting stripes in the paradise whiptail, *Pentapodus paradiseus*. J. Exp. Biol. 206, 3607–3613.
McGraw, K.J., Toomey, M.B., Nolan, P.M., Morehouse, N.I., Massaro, M., Jouventin, P., 2007. A description of unique fluorescent yellow pigments in penguin feathers. Pigment Cell Res. 20, 301–304.
Michielsen, K., Stavenga, D.G., 2008. Gyroid cuticular structures in butterfly wing scales: biological photonic crystals. J. R. Soc. Interface 5, 85–94.
Michielsen, K., De Raedt, H., Stavenga, D.G., 2010. Reflectivity of the gyroid biophotonic crystals in the ventral wing scales of the green hairstreak butterfly, *Callophrys rubi*. J. R. Soc. Interface 7, 765–771.
Mie, G., 1908. Beiträge zur Optik trüber Medien, speziell kolloidaler Metallösungen. Ann. Phys. 330, 377–445.
Mishchenko, M.I., Travis, L.D., Mackowski, D.W., 1996. T-matrix computations of light scattering by nonspherical particles: a review. J. Quant. Spectrosc. Radiat. Transf. 55, 535–575.
Moharam, M.G., Gaylord, T.K., 1981. Rigorous coupled-wave analysis of planar-grating diffraction. J. Opt. Soc. Am. 71, 811–818.
Morehouse, N.I., Vukusic, P., Rutowski, R., 2007. Pterin pigment granules are responsible for both broadband light scattering and wavelength selective absorption in the wing scales of pierid butterflies. Proc. R. Soc. B 274, 359–366.
Morris, R.B., 1975. Iridescence from diffraction structures in the wing scales of *Callophrys rubi*, the green hairstreak. J. Ent. Ser. A 49, 149–154.
Mouchet, S.R., Deparis, O., Vigneron, J.-P., 2012a. Unexplained high sensitivity of the reflectance of porous natural photonic structures to the presence of gases and vapours in the atmosphere. Proc. SPIE 8424, 842425.

Mouchet, S.R., Vigneron, J.-P., Colomer, J.-F., Vandenbem, C., Deparis, O., 2012b. Additive photonic colors in the Brazilian diamond weevil, *Entimus imperialis*. Proc. SPIE 8480, 848003.
Mouchet, S.R., Colomer, J.-F., Vandenbem, C., Deparis, O., Vigneron, J.-P., 2013. Method for modeling additive color effect in photonic polycrystals with form anisotropic elements: the case of *Entimus imperialis* weevil. Opt. Express 21, 13228–13240.
Mouchet, S.R., Lobet, M., Kolaric, B., Kaczmarek, A.M., Van Deun, R., Vukusic, P., Deparis, O., Van Hooijdonk, E., 2016a. Controlled fluorescence in a beetle's photonic structure and its sensitivity to environmentally induced changes. Proc. R. Soc. B 283, 20162334.
Mouchet, S.R., Tabarrant, T., Lucas, S., Su, B.-L., Vukusic, P., Deparis, O., 2016b. Vapor sensing with a natural photonic cell. Opt. Express 24, 12267–12280.
Mouchet, S.R., Van Hooijdonk, E., Welch, V.L., Louette, P., Colomer, J.-F., Su, B.-L., Deparis, O., 2016c. Liquid-induced colour change in a beetle: the concept of a photonic cell. Sci. Rep. 6, 19322.
Mouchet, S.R., Lobet, M., Kolaric, B., Kaczmarek, A.M., Van Deun, R., Vukusic, P., Deparis, O., Van Hooijdonk, E., 2017a. Photonic scales of *Hoplia coerulea* beetle: any colour you like. Mater. Today Proc. 4, 4979–4986.
Mouchet, S.R., Van Hooijdonk, E., Welch, V.L., Louette, P., Tabarrant, T., Vukusic, P., Lucas, S., Colomer, J.-F., Su, B.-L., Deparis, O., 2017b. Assessment of environmental spectral ellipsometry for characterising fluid-induced colour changes in natural photonic structures. Mater. Today Proc. 4, 4987–4997.
Obara, Y., 1970. Studies on the mating behavior of the white cabbage butterfly, *Pieris rapae crucivora* Boisduval. Z. Vgl. Physiol. 69, 99–116.
Oxford, G.S., Gillespie, R.G., 1998. Evolution and ecology of spider coloration. Annu. Rev. Entomol. 43, 619–643.
Parker, A.R., 2000. 515 million years of structural colour. J. Opt. A Pure Appl. Opt. 2, R15–R28.
Parker, A.R., 2005. A geological history of reflecting optics. J. R. Soc. Interface 2, 1–17.
Pavan, M., Vachon, M., 1954. Sur l'existence d'une substance fluorescente dans les téguments des Scorpions (Arachnides). C. R. Acad. Sci. 239, 1700–1702.
Phillips, C.E.S., 1927. Fluorescence of sea anemones. Nature 119, 747.
Piszter, G., Kertész, K., Vértesy, Z., Bálint, Z., Biró, L.P., 2014. Substance specific chemical sensing with pristine and modified photonic nanoarchitectures occurring in blue butterfly wing scales. Opt. Express 22, 22649–22660.
Poncelet, O., Tallier, G., Mouchet, S.R., Crahay, A., Rasson, J., Kotipalli, R., Deparis, O., Francis, L.A., 2016. Vapour sensitivity of an ALD hierarchical photonic structure inspired by *Morpho*. Bioinspir. Biomim. 11, 036011.
Potyrailo, R.A., Naik, R.R., 2013. Bionanomaterials and bioinspired nanostructures for selective vapor sensing. Annu. Rev. Mater. Res. 43, 307–334.
Potyrailo, R.A., Ghiradella, H., Vertiatchikh, A., Dovidenko, K., Cournoyer, J.R., Olson, E., 2007. *Morpho* butterfly wing scales demonstrate highly selective vapour response. Nat. Photonics 1, 123–128.
Potyrailo, R.A., Starkey, T.A., Vukusic, P., Ghiradella, H., Vasudev, M., Bunning, T., Naik, R.R., Tang, Z., Larsen, M., Deng, T., Zhong, S., Palacios, M., Grande, J.C., Zorn, G., Goddard, G., Zalubovsky, S., 2013. Discovery of the surface polarity gradient on iridescent *Morpho* butterfly scales reveals a mechanism of their selective vapor response. Proc. Natl. Acad. Sci. U. S. A. 110, 15567–15572.
Pouya, C., Vukusic, P., 2012. Electromagnetic characterization of millimetre-scale replicas of the gyroid photonic crystal found in the butterfly *Parides sesostris*. Interface Focus 2, 645–650.

Pouya, C., Stavenga, D.G., Vukusic, P., 2011. Discovery of ordered and quasi-ordered photonic crystal structures in the scales of the beetle *Eupholus magnificus*. Opt. Express 19, 11355–11364.

Pris, A.D., Utturkar, Y., Surman, C., Morris, W.G., Vert, A., Zalyubovskiy, S., Deng, T., Ghiradella, H.T., Potyrailo, R.A., 2012. Towards high-speed imaging of infrared photons with bio-inspired nanoarchitectures. Nat. Photonics 6, 195–200.

Prum, R.O., Torres, R.H., 2004. Structural colouration of mammalian skin: convergent evolution of coherently scattering dermal collagen arrays. J. Exp. Biol. 207, 2157–2172.

Prum, R.O., Torres, R.H., Williamson, S., Dyck, J., 1998. Coherent light scattering by blue feather barbs. Nature 396, 28–29.

Prum, R.O., Quinn, T., Torres, R.H., 2006. Anatomically diverse butterfly scales all produce structural colours by coherent scattering. J. Exp. Biol. 209, 748–765.

Rassart, M., Simonis, P., Bay, A., Deparis, O., Vigneron, J.-P., 2009. Scale coloration change following water absorption in the beetle *Hoplia coerulea* (Coleoptera). Phys. Rev. E Stat. Nonlin. Soft Matter Phys. 80 . 031910.

Rasson, J., Poncelet, O., Mouchet, S.R., Deparis, O., Francis, L.A., 2017. Vapor sensing using a bio-inspired porous silicon photonic crystal. Mater. Today Proc. 4, 5006–5012.

Rayleigh, J.W.S., 1887. XVII. On the maintenance of vibrations by forces of double frequency and on the propagation of waves through a medium endowed with a periodic structure. Philos. Mag. 24, 145–159.

Rayleigh, J.W.S., 1888. XXVI. On the remarkable phenomenon of crystalline reflexion described by Prof. Stokes. Philos. Mag. 26, 256–265. S5.

Saranathan, V., Osuji, C.O., Mochrie, S.G.J., Noh, H., Narayanan, S., Sandy, A., Dufresne, E.R., Prum, R.O., 2010. Structure, function, and self-assembly of single network gyroid ($I4_132$) photonic crystals in butterfly wing scales. Proc. Natl. Acad. Sci. U. S. A. 107, 11676–11681.

Seago, A.E., Brady, P., Vigneron, J.-P., Schultz, T.D., 2009. Gold bugs and beyond: a review of iridescence and structural colour mechanisms in beetles (Coleoptera). J. R. Soc. Interface 6, S165–S184.

Shi, N.N., Tsai, C.C., Camino, F., Bernard, G.D., Yu, N., Wehner, R., 2015. Keeping cool: enhanced optical reflection and radiative heat dissipation in Saharan silver ants. Science 349, 298–301.

Siddique, R.H., Gomard, G., Hölscher, H., 2015. The role of random nanostructures for the omnidirectional anti-reflection properties of the glasswing butterfly. Nat. Commun. 6, 6909.

Siddique, R.H., Vignolini, S., Bartels, C., Wacker, I., Hölscher, H., 2016. Colour formation on the wings of the butterfly *Hypolimnas salmacis* by scale stacking. Sci. Rep. 6, 36204.

Silberglied, R., 1984. Visual communication and sexual selection among butteries. In: Vane-Wright, R.I., Ackery, P.E. (Eds.), In: The Biology of Butteries, vol 11. Academic Press, London, pp. 207–223.

Sollas, I.B.J., 1907. On the identification of chitin by its physical constants. Proc. R. Soc. B 79, 474–481.

Stavenga, D.G., 2014. Thin film and multilayer optics cause structural colors of many insects and birds. Mater. Today Proc. 1S, 109–121.

Stavenga, D.G., Stowe, S., Siebke, K., Zeil, J., Arikawa, K., 2004. Butterfly wing colours: scale beads make white pierid wings brighter. Proc. R. Soc. B 271, 1577–1584.

Stavenga, D.G., Giraldo, M.A., Hoenders, B.J., 2006. Reflectance and transmittance of light scattering scales stacked on the wings of pierid butterflies. Opt. Express 14, 4880–4890.

Stavenga, D.G., Giraldo, M.A., Leertouwer, H.L., 2010. Butterfly wing colors: glass scales of *Graphium sarpedon* cause polarized iridescence and enhance blue/green pigment coloration of the wing membrane. J. Exp. Biol. 213, 1731–1739.

Stavenga, D.G., Matsushita, A., Arikawa, K., Leertouwer, H.L., Wilts, B.D., 2012. Glass scales on the wing of the swordtail butterfly *Graphium sarpedon* act as thin film polarizing reflectors. J. Exp. Biol. 215, 657–662.

Stavenga, D.G., Leertouwer, H.L., Wilts, B.D., 2014. Coloration principles of nymphaline butterflies—thin films, melanin, ommochromes and wing scale stacking. J. Exp. Biol. 217, 2171–2180.

Taboada, C., Brunetti, A.E., Pedron, F.N., Carnevale Neto, F., Estrin, D.A., Bari, S.E., Chemes, L.B., Peporine Lopes, N., Lagorio, M.G., Faivovich, J., 2017. Naturally occurring fluorescence in frogs. Proc. Natl. Acad. Sci. U. S. A. 114, 3672–3677.

Taflove, A., Hagness, S.C., 2005. Computational Electrodynamics: The Finite-Difference Time-Domain Method, third ed. Artech House, Norwood.

Tamáska, I., Kertész, K., Vértesy, Z., Bálint, Z., Kun, A., Yen, S.-H., Biró, L.P., 2013. Color changes upon cooling of Lepidoptera scales containing photonic nanoarchitectures, and a method for identifying the changes. J. Insect Sci. 13, 1–15.

Tan, T.L., Wong, D., Lee, P., 2004. Iridescence of a shell of mollusk *Haliotis glabra*. Opt. Express 12, 4847–4854.

Tani, K., Watari, F., Uo, M., Morita, M., 2004. Fluorescent properties of porcelain-restored teeth and their discrimination. Mater. Trans. 45, 1010–1014.

Trzeciak, T.M., Wilts, B.D., Stavenga, D.G., Vukusic, P., 2012. Variable multilayer reflection together with long-pass filtering pigment determines the wing coloration of papilionid butterflies of the *nireus* group. Opt. Express 20, 8877–8890.

Van Hooijdonk, E., Barthou, C., Vigneron, J.-P., Berthier, S., 2011. Detailed experimental analysis of the structural fluorescence in the butterfly *Morpho sulkowskyi* (Nymphalidae). J. Nanophotonics 5, 053525.

Van Hooijdonk, E., Barthou, C., Vigneron, J.-P., Berthier, S., 2012a. Angular dependence of structural fluorescent emission from the scales of the male butterfly *Troïdes magellanus* (Papilionidae). J. Opt. Soc. Am. B 29, 1104–1111.

Van Hooijdonk, E., Berthier, S., Vigneron, J.-P., 2012b. Bio-inspired approach of the fluorescence emission properties in the scarabaeid beetle *Hoplia coerulea* (Coleoptera): modeling by transfer matrix optical simulations. J. Appl. Phys. 112, 114702.

Van Hooijdonk, E., Berthier, S., Vigneron, J.-P., 2012c. Contribution of both the upperside and the underside of the wing on the iridescence in the male butterfly *Troïdes magellanus* (Papilionidae). J. Appl. Phys. 112, 074702.

Van Hooijdonk, E., Barthou, C., Vigneron, J.-P., Berthier, S., 2013. Yellow structurally modified fluorescence in the longhorn beetles *Celosterna pollinosa sulfurea* and *Phosphorus virescens* (Cerambycidae). J. Lumin. 136, 313–321.

Vigneron, J.-P., Lousse, V., 2006. Variation of a photonic crystal color with the Miller indices of the exposed surface. Proc. SPIE 6128, 61281G.

Vigneron, J.-P., Simonis, P., 2010. Structural colours. Adv. Insect Physiol. 38, 181–218.

Vigneron, J.-P., Colomer, J.-F., Vigneron, N., Lousse, V., 2005a. Natural layer-by-layer photonic structure in the squamae of *Hoplia coerulea* (Coleoptera). Phys. Rev. E Stat. Nonlin. Soft Matter Phys. 72, 061904.

Vigneron, J.-P., Lousse, V., Biró, L.P., Vértesy, Z., Bálint, Z., 2005b. Reflectance of topologically disordered photonic-crystal films. Proc. SPIE 5733, 308–315.

Vigneron, J.-P., Colomer, J.-F., Rassart, M., Ingram, A.I., Lousse, V., 2006. Structural origin of the colored reflections from the black-billed magpie feathers. Phys. Rev. E Stat. Nonlin. Soft Matter Phys. 73, 021914.

Vigneron, J.-P., Kertész, K., Vértesy, Z., Rassart, M., Lousse, V., Bálint, Z., Biró, L.P., 2008. Correlated diffraction and fluorescence in the backscattering iridescence of the male butterfly *Troides magellanus* (Papilionidae). Phys. Rev. E Stat. Nonlin. Soft Matter Phys. 78, 021903.

Vigneron, J.-P., Simonis, P., Aiello, A., Bay, A., Windsor, D.M., Colomer, J.-F., Rassart, M., 2010. Reverse color sequence in the diffraction of white light by the wing of the male butterfly *Pierella luna* (Nymphalidae: Satyrinae). Phys. Rev. E Stat. Nonlin. Soft Matter Phys. 82, 021903.

Vignolini, S., Rudall, P.J., Rowland, A.V., Reed, A., Moyroud, E., Faden, R.B., Baumberg, J.J., Glover, B.J., Steiner, U., 2012. Pointillist structural color in *Pollia* fruit. Proc. Natl. Acad. Sci. U. S. A. 109, 15712–15715.

Vukusic, P., 2005. Structural colour effects in Lepidoptera. In: Kinoshita, S., Yoshioka, S. (Eds.), Structural Colors in Biological Systems: Principles and Applications. Osaka University Press, Osaka, pp. 95–112.

Vukusic, P., Hooper, I., 2005. Directionally controlled fluorescence emission in butterflies. Science 310, 1151.

Vukusic, P., Sambles, J.R., 2001. Shedding light on butterfly wings. Proc. SPIE 4438, 85–95.

Vukusic, P., Sambles, J.R., 2003. Photonic structures in biology. Nature 424, 852–855.

Vukusic, P., Sambles, J.R., Lawrence, C.R., Wootton, R.J., 1999. Quantified interference and diffraction in single *Morpho* butterfly scales. Proc. R. Soc. B 266, 1403–1411.

Vukusic, P., Sambles, J.R., Ghiradella, H., 2000a. Optical classification of microstructure in butterfly wing-scales. Photon. Sci. News 6, 61–66.

Vukusic, P., Sambles, J.R., Lawrence, C.R., 2000b. Structural colour: colour mixing in wing scales of a butterfly. Nature 404, 457.

Vukusic, P., Sambles, J.R., Lawrence, C.R., Wakely, G., 2001a. Sculpted-multilayer optical effects in two species of *Papilio* butterfly. Appl. Opt. 40, 1116–1125.

Vukusic, P., Sambles, J.R., Lawrence, C.R., Wootton, R.J., 2001b. Structural colour: now you see it—now you don't. Nature 410, 36.

Vukusic, P., Sambles, J.R., Lawrence, C.R., Wootton, R.J., 2002. Limited-view iridescence in the butterfly *Ancyluris meliboeus*. Proc. R. Soc. B 269, 7–14.

Vukusic, P., Sambles, J.R., Lawrence, C.R., 2004. Structurally assisted blackness in butterfly scales. Proc. R. Soc. B 271, S237–S239.

Vukusic, P., Hallam, B., Noyes, J., 2007. Brilliant whiteness in ultrathin beetle scales. Science 315, 348.

Vukusic, P., Kelly, R., Hooper, I., 2009. A biological sub-micron thickness optical broadband reflector characterized using both light and microwaves. J. R. Soc. Interface 6, S193–S201.

Wang, W., Zhang, W., Fang, X., Huang, Y., Liu, Q., Gu, J., Zhang, D., 2014. Demonstration of higher colour response with ambient refractive index in *Papilio blumei* as compared to *Morpho rhetenor*. Sci. Rep. 4, 5591.

Welch, V.L., Lousse, V., Deparis, O., Parker, A.R., Vigneron, J.-P., 2007. Orange reflection from a three-dimensional photonic crystal in the scales of the weevil *Pachyrrhynchus congestus pavonius* (Curculionidae). Phys. Rev. E Stat. Nonlin. Soft Matter Phys. 75, 041919.

Welch, V.L., Van Hooijdonk, E., Intrater, N., Vigneron, J.-P., 2012. Fluorescence in insects. Proc. SPIE 8480, 848004.

Wilts, B.D., Leertouwer, H.L., Stavenga, D.G., 2009. Imaging scatterometry and microspectrophotometry of lycaenid butterfly wing scales with perforated multilayers. J. R. Soc. Interface 6, S185–S192.

Wilts, B.D., Michielsen, K., De Raedt, H., Stavenga, D.G., 2012a. Hemispherical Brillouin zone imaging of a diamond-type biological photonic crystal. J. R. Soc. Interface 9, 1609–1614.

Wilts, B.D., Michielsen, K., Kuipers, J., De Raedt, H., Stavenga, D.G., 2012b. Brilliant camouflage: photonic crystals in the diamond weevil, *Entimus imperialis*. Proc. R. Soc. B 279, 2524–2530.

Wilts, B.D., Trzeciak, T.M., Vukusic, P., Stavenga, D.G., 2012c. Papiliochrome II pigment reduces the angle dependency of structural wing colouration in *nireus* group papilionids. J. Exp. Biol. 215, 793–805.

Wu, X., Erbe, A., Raabe, D., Fabritius, H.-O., 2013. Extreme optical properties tuned through phase substitution in a structurally optimized biological photonic polycrystal. Adv. Funct. Mater. 23, 3615–3620.

Yablonovitch, E., 1987. Inhibited spontaneous emission in solid-state physics and electronics. Phys. Rev. Lett. 58, 2059–2062.

Yagi, N., 1954. Note of electron microscope research on pterin pigment in the scales of pierid butterflies. Annot. Zool. Jpn. 27, 113–114.

Yang, X., Peng, Z., Zuo, H., Shi, T., Liao, G., 2011. Using hierarchy architecture of *Morpho* butterfly scales for chemical sensing: experiment and modeling. Sens. Actuators A Phys. 167, 367–373.

Yee, K.S., 1966. Numerical solution of initial boundary value problems involving Maxwell's equations in isotropic media. IEEE Trans. Antennas Propag. 14, 302–307.

Yeh, P., 2005. Optical Waves in Layered Media, second ed. Wiley-Interscience, Hoboken.

Yin, H., Shi, L., Sha, J., Li, Y., Qin, Y., Dong, B., Meyer, S., Liu, X., Zhao, L., Zi, J., 2006. Iridescence in the neck feathers of domestic pigeons. Phys. Rev. E Stat. Nonlin. Soft Matter Phys. 74, 051916.

Yoshida, A., 2002. Antireflection of the butterfly and moth wings through microstructure. Forma 17, 75–89.

Yoshida, A., Motoyama, M., Kosaku, A., Miyamoto, K., 1996. Nanoprotuberance array in the transparent wing of a hawkmoth, *Cephonodes hylas*. Zoolog. Sci. 13, 525–526.

Yoshida, A., Motoyama, M., Kosaku, A., Miyamoto, K., 1997. Antireflective nanoprotuberance array in the transparent wing of a hawkmoth, *Cephonodes hylas*. Zoolog. Sci. 14, 737–741.

Yoshioka, S., Kinoshita, S., 2002. Effect of macroscopic structure in iridescent color of the peacock feathers. Forma 17, 169–181.

Yoshioka, S., Kinoshita, S., 2004. Wavelength-selective and anisotropic light-diffusing scale on the wing of the *Morpho* butterfly. Proc. R. Soc. B 271, 581–587.

Yoshioka, S., Kinoshita, S., 2006a. Single-scale spectroscopy of structurally colored butterflies: measurements of quantified reflectance and transmittance. J. Opt. Soc. Am. A 23, 134–141.

Yoshioka, S., Kinoshita, S., 2006b. Structural or pigmentary? Origin of the distinctive white stripe on the blue wing of a *Morpho* butterfly. Proc. R. Soc. B 273, 129–134.

Yoshioka, S., Kinoshita, S., 2007. Polarization-sensitive color mixing in the wing of the Madagascan sunset moth. Opt. Express 15, 2691–2701.

Yoshioka, S., Kinoshita, S., 2011. Direct determination of the refractive index of natural multilayer systems. Phys. Rev. E Stat. Nonlin. Soft Matter Phys. 83, 051917.

Zang, X., Tan, Y., Lv, Z., Gu, J., Zhang, D., 2012. Moth wing scales as optical pH sensors. Sens. Actuators B 166-167, 824–828.

CHAPTER TWO

More Than Colours: An Eco-Evolutionary Framework for Wing Shape Diversity in Butterflies

Dirleane O. Rossato*, Lucas A. Kaminski*, Cristiano A. Iserhard†, Leandro Duarte*
*Universidade Federal do Rio Grande do Sul (UFRGS), Porto Alegre, Rio Grande do Sul, Brazil
†Universidade Federal de Pelotas (UFPEL), Pelotas, Rio Grande do Sul, Brazil

Contents

Abstract

Wings are key innovations that have revolutionized the life of winged organisms and have moulded terrestrial ecosystems through new interactions (e.g. pollination). Among flying organisms, butterflies draw attention for their astonishing diversity of colour patterns, but their wings are much more than colours. The wing morphology may vary for different reasons, including communication. Thus, the different factors, acting together or under conflicting ways, moulded the wing shape. Mimicry rings are promising model systems for evaluating the evolution of wing shapes since trait convergence for mimicry is an identifiable force among others selective pressures. An eco-evolutionary framework can help us disentangle the relative role of each step in the evolution of wing shapes. Biotic and abiotic filters can affect wing morphology through pathway changes during the ontogeny, including sexual dimorphism, developmental trade-offs,

Advances in Insect Physiology, Volume 54
ISSN 0065-2806
https://doi.org/10.1016/bs.aiip.2017.11.003

nutritional food quality and environmental conditions. During the adult stage, both sexual and natural selection determine the combination of genes that are selected and will persist for future generations. All these selective pressures acting repeatedly over the time must result in different macroevolutionary scenarios of traits, like as genetic drift, vicariance, adaptive radiation and speciation. Finally, we propose that comparative studies on mimicry rings, with closely related species presenting changes in wing morphology according to both environmental and sexual behaviour, including sex-limited dimorphism, can be useful systems to understand the role of selective pressures driving the evolution of wing shapes.

1. INTRODUCTION

Currently, the ability to fly is present only in insects, birds and bats. It is an extraordinary trait that has revolutionized terrestrial ecosystems, encouraging new, wondrous interactions, including, for example, specialized pollination systems (Regal, 1977). The wing is the key innovation for winged organisms that allows for travel through open-air offering solutions for many organismal demands (Alexander, 2015). Movement through the air favoured the exploration within and among different habitats, the search and exploration of new adaptive zones, the expansion of the distribution range across large geographical regions and the transposition of high elevations and to fly though long distance between seasons (e.g. DeVries et al., 2010; Greenslade et al., 1999; Johnson, 1969). Flight-related traits should be important in determining fitness, as the ability to take flight improved foraging efficiency, territorial defence, mate search, escape of predation and avoidance of inadequate conditions (Grimaldi and Engel, 2005; Norberg, 1990). Besides the primary flying function, the wing can operate with communication signals, acting in sexual and territorial displays, in cryptic (camouflage) or warning defence (Batesian and Müllerian mimicry), in thermoregulation and other behaviours (e.g. Baker, 1972; Betts and Wootton, 1988; Davies, 1978; De Keyser et al., 2015; Kingsolver, 1985; Papageorgis, 1975). Studies underlying different functional aspects in wings have allowed for a greater understanding of both the history and the selective pressures throughout the evolution of winged animals. In this review, we will use the concept of functional traits from Violle et al. (2007) in which the trait could be what impacts fitness indirectly on growth, reproduction and survival. Wing size and shape, for example, result from interactions between adaptive process, phylogenetic history, developmental constraints, as well as on the macroevolution scale (Bai et al., 2012; Chazot et al., 2016; Grimaldi and Engel, 2005). Thus, studies that look for the relationships between trait functionality and adaptation under micro- and macroevolutionary scales may allow us to understand different selective pressures and constraints on ecomorphological traits (Fig. 1). Simultaneously, we could come closer to learning the true history of wing shape variation as a result of different processes across time.

One promising winged group that could help to evaluate micro- and macroevolutionary process are the butterflies. As noted by Bates (1864), the relevance of these insects for Biological Science would one day be recognized by the adaptive messages "written" on the expanded wing surfaces of butterflies. Surely, Bates wrote this in the context of his recent discoveries on the adaptive significance of wing colour pattern convergence—Batesian mimicry (Bates, 1862). Today, butterflies are model organisms in evolutionary biology, including genetics, physiology, development and ecology (Beldade and Brakefield, 2002; Kronforst and Papa, 2015). This group is closely related with the environment where it lives, having a high association with biotic and abiotic resources and factors at different scales, from microhabitats to large landscapes (Bonebrake et al., 2010; Brown, 1997; Freitas and Marini-Filho, 2011). Therefore, butterflies could respond to natural or anthropogenic changes in a short period of time and thereby help us to understand eco-evolutionary processes, such as different spatial scales, time, intensity and frequency of these influences. The selective pressure on butterflies can affect different aspects. Thus, they also can be used to explain the effects of these changes on growth, reproduction and survival. Wing shape is one of these remarkable traits that can uncover the history of the sexual selection, dispersion, flight type, speciation and diversification. All of these processes can evolve in conflict with morphological convergence to mimicry rings (Srygley, 1994). The micro- and macroevolutionary processes could be understood by looking for DNA sequences; however, in microevolution the change occurs with gene frequency within populations and over short periods of time. On the other hand, for macroevolution, the objective is at the species level, and in order to understand this, we need look to the tree of the related species accessing the diversity of the clade (Fig. 1). Using phylogenetic methods, we can investigate how wing traits change over time considering how geographic or other ecological factors (microhabitat use and sexual selection) and macroevolutionary processes (shared ancestry) act on wing evolution.

Recently, there is an increasing in studies on macroevolution approach due to the advent of robust and almost complete phylogenetic hypothesis based on several marks and molecular clocks, calibrated with fossils for dating with confidence (Cadotte and Davies, 2016). These advances have made it

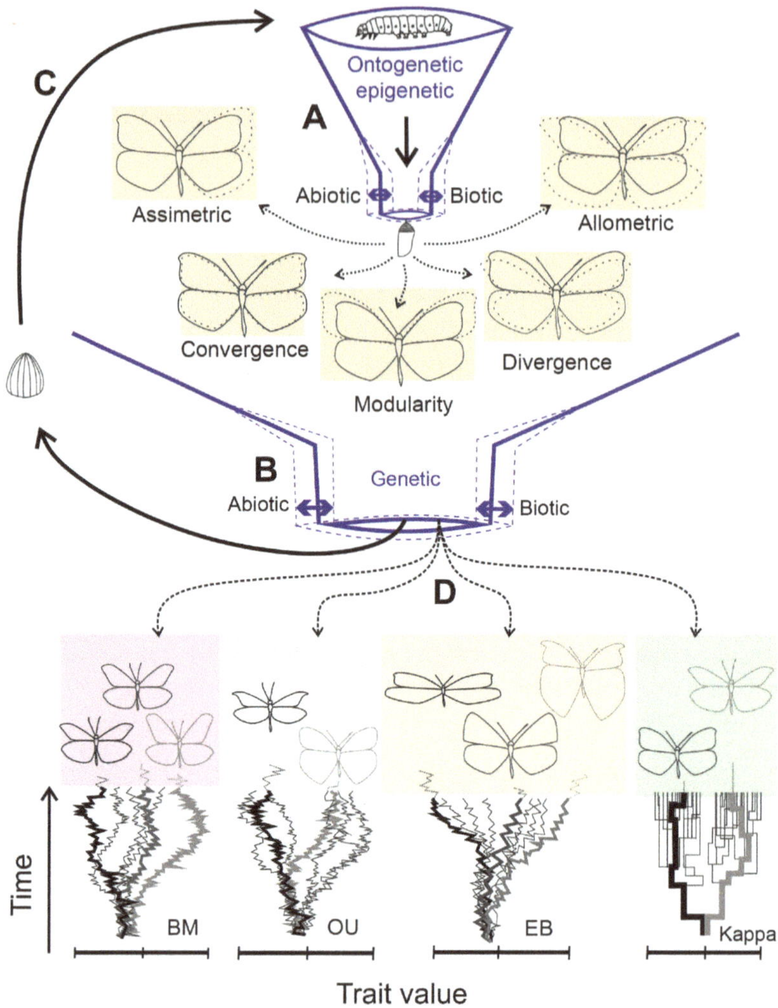

Fig. 1 The eco-evolutionary framework depicted micro- and macroevolutionary processes acting on wing shape diversity in butterflies. At the top, the selective pressures that could act on a small scale of individuals level are represented (steps A, B and C), which with much repeated time could result in macroevolutionary scenarios described by distinct models of evolution (step D). Step 1A starts in the immature stages, which are affected by biotic and abiotic conditions through ontogeny and is the main stage for the epigenetic controls. *Arrow* inside the filter represents the ontogenetic framework being molded by biotic (e.g. predation) and abiotic (e.g. food quality, environmental conditions) filters. The interaction of these external factors with individuals features may influence the gene expression, also by the epigenetic control. *Dotted arrow*, after to pupal stage, shows that similar genotypes could result in distinct wing shapes (asymmetry, convergence, modularity, divergence, allometry) depending of the power of the selective forces. Step B starts with adult genetic information to be selected for biotic (e.g. predation) and abiotic (e.g. habitat type) factors, this step ends at the egg stage as result of mating with genetic information selected and survival to predation. Step C is the restart of this cycle of the selections acting on individuals. Step D starts the macroevolutionary process, a consequence of several recurrent cycles of microevolutionary steps (A, B and C) through evolutionary time of lineages. *Dotted arrow* shows the possibility of distinct evolution models occur, each one reflects different evolutionary hypotheses on the time (*BM*, Brownian model; *OU*, Ornstein–Uhlenbeck; *EB*, early-burst; kappa).

possible to understand the time and mode of evolution for several living species/lineages. This emergence of analytic tools makes it possible to evaluate how morphological traits evolved and propose some selective pressure that could have acted under it. In this view, it is important to consider the different functions associated with both the traits, the historic and developmental constraints. Integration of these approaches can clarify how forces operate upon the evolution of traits. Seeing how selective pressures affect the wing shape variation on the mimicry ring contexts was described by Srygley (1994) as one possible solution for finding evidence of convergence on wing shape between comimetics, separating one from the other effects. Moreover, it is important to choose appropriate tools, since the non-convergence in the shape of the wing (Srygley, 1994), may be the result of inadequate tools. For butterflies, this is an emerging avenue of research, and many recently published papers addressing the evolution of wing shapes and its relation to environmental transitions and mimicry (e.g. Cespedes et al., 2015; Chazot et al., 2016; DeVries et al., 2010; Jones et al., 2013; Mérot et al., 2016; Rossato et al., in press). In this review, we present the recent advances in the study of wing shapes and discuss the different aspects that act on wing morphology in butterflies, taking into account: (i) ontogenetic drivers (Fig. 1, step A); (ii) impact of the biotic and abiotic filters (Fig. 1, steps A–C); and (iii) the macroevolutionary process (Fig. 1, step D). Finally, we present an eco-evolutionary framework identifying what is potentially driving the evolution of butterfly wing shape diversity.

2. WING SHAPE AND FUNCTION

The primary function of the wing is locomotion. In this sense, much of the wing shape studies are focused on flight physics, material properties and aerodynamics (revised in Wootton, 1992, 2002). Affecting drag or lift, and thus flight behaviour, the wing size and shape are strongly related to physical wing properties (Berwaerts et al., 2002; Chazot et al., 2016; Dudley, 2000; Wootton, 1992). Moreover, wing shape is related to many other ecological functions, being a sound predictor of functional adaptation. Thus, wing shape can be a good proxy of different processes and can be used to answer an array of questions in winged organisms (Johansson et al., 2009).

2.1 Evaluating the Wing Shape

Usually wing shape information is widely evaluated by means of the aspect ratio index—proportion between length and wing area (Betts and Wootton, 1988;

Bots et al., 2012; Hassall, 2015; Johansson et al., 2009), or through the use of the elliptical Fourier (Ferson et al., 1985; Holwell and Herberstein, 2010; Jones et al., 2013). However, the best way to obtain the shape is by removing the size, rotation and translation information (Bookstein, 1991), though of generalized procrustes analysis (GPA) (Rohlf and Slice, 1990) in geometric morphometrics (GM) (Adams et al., 2004). GM uses a Cartesian landmark coordinates to obtain the shape of the wing set, solving the problems of linear measurement when the same set of linear measurements can be obtained from different shapes (Van der Molen et al., 2007). This technique is widely used in many research areas in biology, including ecology, evolution, development, zoology, botany, taxonomy and even palaeontology.

GM analysis revolutionized the way that we study the shape morphology in many organisms, particularly in butterflies. This emergence increases the knowledge to a better understanding of the different selective pressures acting on morphological traits that can be a result of evolutionary history and ecological processes. Thus, for butterflies this method is crucial, since the evaluation of phenotypic plasticity of the wing shape of *Pararge aegeria* (Breuker et al., 2010), to study biogeographic scenario including cryptic species combined with genetic markers to explain the relationship between historical dispersal and morphological variation (Dincă et al., 2011). In mimetic *Heliconius* butterflies, Jorge et al. (2011) evaluated the wing variation of *Heliconius erato* rearing on distinct host plant species, and Jones et al. (2013) evaluated the wing shape variation and convergence in a complex of mimicry butterflies. Also, Mérot et al. (2016) evaluated the phenotypic resemblance between comimics in different localities with varying species composition. Finally, Rossato et al. (in press) evaluated the resemblance of the wing and red band between the spatial cooccurrence of comimetic species.

2.1.1 Allometry

The morphological variation in organism bodies is integrated (Olson and Miller, 1958). The study of the different growth rates between parts of the individual and the corresponding impact of these changes under the perspective of ecology and evolution is known as allometry (Huxley and Teissier, 1936). Allometric variation can be evaluated at different levels, between species of the same population (static), between different stages of development (ontogenetic) and throughout the evolutionary history (phylogenetic) (see Cock, 1966; Klingenberg and Zimmermann, 1992;

Moreira et al., 2017). Rossato et al. (in press) used the allometric index to understand differences between the red band and wing size in males and females in some butterfly species in a mimetic ring. Furthermore, we can also call allometry the relationship between size and shape since these parts must vary in a coordinated manner, change in size will result change in shape, so that each organism forms a completely functioning whole (Cheverud, 1996; Olson and Miller, 1958). The GPA of GM removes the isometric effect, but not the allometric relationships. Thus, this approach does not consider the changes in shape when the size has changed. In a critical review, Outomuro and Johansson (2017) compared different studies on GM analysis and the relationship between shape and size, and found that size explains 88% of shape variation, suggesting the importance in incorporating the size in order to understand shape variation. In addition, Outomuro et al. (2013) studying wing shape allometry and aerodynamics in calopterygid damselflies suggest that the relationship between wing shape and size is taxa-specific and differs among closely related lineages.

Still, to analyse wing shape and the factors that impact it, the allometry is probably the best way to assess the potential contribution of size on the shape variation. This inclusion will not only be more informative and will enable a better understanding of the relationship between shape and size (Outomuro and Johansson, 2017), but also the variation on the growth rate between different parts of butterflies related to different selective pressures.

2.1.2 Modularity

Another important aspect of the wing shape is the modularity: the idea that some parts or regions (modules) of a structure are internally integrated by interactions among traits but are relatively independent from other modules (Klingenberg, 2008). The role of developmental and genetic integration in evolution is contentious. One hypothesis states that integration acts as a constraint on evolution, whereas an alternative is that developmental and genetic systems evolve to match the functional modularity of organisms (Klingenberg et al., 2010). This question was studied by Klingenberg et al. (2010) who examined morphological structures using GM in the cricket wing where developmental and functional modules are discordant, making it possible to distinguish between the two alternatives. The patterns of genetic, phenotypic and developmental integration were clearly similar, but not identical. In this sense, the hypothesis that genetic and developmental integration evolve to match functional modularity can be rejected (Klingenberg et al., 2010), on this case.

Fore and hind wings originate from distinct imaginal discs, so it is expected that wings are developmentally autonomous. These two pairs of wings constitute two developmental modules in which groups of traits evolve relatively independently of each other (Breuker et al., 2006), that is, different wings can play different functions in flight (Grodnitsky et al., 1994). To understand the integration of wings and the extent to which it relates to ecology is a fundamental question for understanding the evolution of butterfly wings. We can assume different functions during flight may have led to a different pattern of wing size and drive diversification (Klingenberg et al., 2001). It is known that forewings drive flight in butterflies (Dudley, 2000) and when the hind wings are removed, the butterfly is still able to fly, so hind wings do not contribute to normal flight (Jantzen and Eisner, 2008), suggesting that hind wings can be related to increase flight speed and manoeuvrability. Therefore, Chazot et al. (2016), while evaluating shapes using morphometric tools, found that both fore and hind wings were similarly affected by microhabitat, both were more elongated in canopy and more narrow rounded in understory, even though they have different involvements in flight. Moreover, they found that wings are more integrated rather than modular, showing similar functionality. The combination of fore and hind wings can change in different microhabitats, and this disappears when the phylogeny effected is controlled. The covariation of fore and hind wings is the result of functional constraints driven by microhabitat; it cannot be distinguished from a neutral phylogenetic signal (Chazot et al., 2016). Focusing on the degree of modularity of different regions of the wings and across fore and hind wings at various levels (intra- vs interspecific, Klingenberg, 2009), should provide future insight into the developmental and functional integration of butterfly wings (Allen et al., 2008; Frankino et al., 2007). Additional to this, could be interest to see the modularity into the same wing, for understand how the veins and parts of the wing answer for different environmental conditions, and for different selective pressures, to mimicry butterflies, could getting the aspect of belonging to ring mimic, like the red band shape in *Heliconius* postman, compared with whole wing shape.

2.1.3 Phylogenetic Relationships

The differences in wing shapes analysed in a phylogenetic context can clarify many functional aspects, since the species are not independent entities. Consequently, the observed traits can be biased due to shared ancestry rather than evolutionary differences (Harvey and Pagel, 1991). In other words, wing shape can be a phylogenetic constraint and not moulded by interaction with the environment. Thus, when we intend to understand whether the change in wing shape is a result of the interaction with ecological and evolution processes, it is important to consider the relationship between the phylogeny and traits, since the trait values can have a high phylogenetic signal and not be the result of the selective pressure on the trait, but of shared evolutionary history. Thus, incorporating phylogenetic correction can help us to understand the evolutionary history and ecological processes on trait evolution (Blankers et al., 2012; Magurran, 2004). Looking at the phylogenetic analysis acting upon the community context, the taxa that are more phylogenetically distant in their composition will have greater phylogenetic diversity in relation to the other community that has closely related taxa (Sobral and Cianciaruso, 2012). This diversity of evolutionary scenarios can reflect trait diversity when these traits evolve for drift, without selective pressure.

3. ONTOGENETIC DRIVERS

Understanding the origin of morphological variation is the main goal of evolutionary developmental biology (evo-devo). Derived from a precursor group of 30 cells that invaginates from the embryonic ectoderm, the butterfly wing is a morphological structure with quantitative multivariate change both at intraspecific and interspecific levels (Houle et al., 2003). Specifically, the wing shape can show some parts with strong patterns of covariation (Klingenberg and Zaklan, 2000), whereas others are independent (Weber, 1992). The shape of the wing discs changes throughout their growth phases in a species-specific pattern. These changes, in the aforementioned instance, must be related to changes in cell processes such as proliferation, shape change, movement and death. Additionally, the wing shape could be affected by genes with prominent developmental roles outside the wing, such as the transcription factors *invected*, *engrailed* and *scalloped*, as well *wingless*, *decapentaplegic* and *vestigial* (Baena-Lopez and Garcia-Bellido, 2006; Carreira et al., 2011; Huang and Huang, 2005). In this view, most of the growth and differentiation of the wing occurs during the late-larval, prepupa and pupa stages; and the orientation of division plays a key role in determining organ shape (Gillies and Cabernard, 2011). Therefore, the evolution in the shape of the wing begins with a change in the progressive spatial pattern of the cell division. But, although the differentiation of the vein morphology occurs during the stage of development of the pupa, it is still in the larval stage that the group of cells expresses the vein-specific proteins during the larval phase and its patterns are generally used to characterize the wing shape through GM tools. These stages begin at the molecular level with highly polygenic genetic inheritance making it difficult to attribute a general understanding of variation in wing size and shape to just one gene. Jones et al. (2012, 2013) suggest that the *P* supergene can be involved in the control of the wing shape among morphs in *Heliconius numata*. This genetic information affects the growth and several cellular processes controlled by the patterned distribution of morphogenesis. However, it is difficult to establish a real causal relationship between specific morphogenesis distribution and specific tissue shapes. In this view, about 50% of the transcriptome exhibits changes in expression during the time course of wing development (O'Keefe et al., 2012).

To understand how the wing evolves in butterfly ontogenetic growth, Nijhout et al. (2014) evaluated the growth pattern on the wing, using landmarks, in two distantly related species with a different shape. The authors found the anterior portion of the wing disc was under stress, supporting the idea that cell elongation in this region is due to stretching imposed by growth and elongation of the more posterior portion of the wing. Growth and cell division in lepidopteran wing discs require both ecdysone and insulin signalling (Nijhout and Grunert, 2002; Nijhout et al., 2007). Likewise, the evolution of wing shapes must come about through evolutionary changes in the spatial and temporal pattern of cell division. These changes, in turn, are likely due to the evolution of the spatial arrangement of the hormone response elements that control the forms of cell division and orientation and the mechanical forces (Marinari et al., 2012; Vincent et al., 2013).

The developmental constraints in the caterpillar stage are relevant factors that can affect the adult butterfly wing shape. Caterpillars, generally, have low mobility and generally feed on a single plant during development. The specialized diet of caterpillars is reported to be a limiting factor shaping the distributions patterns in butterflies (Graça et al., 2015) even though adults may be mobile and forage across extensive areas (Marchant et al., 2015). In this view, Jorge et al. (2011) found a variation of size and wing shape among individuals that fed on different plant species, reinforcing the role of plasticity on host–plant use. In fact, the host plant is a paramount environmental factor in the ecology of herbivorous insects, indicating that a large proportion of total morphological variation has an environmental origin (Jorge et al., 2011). Besides that, the biotic factor as the presence of predators can induce the phenotypic plasticity in larvae, causing changes on their morphology without differentiation at the genetic level, inducing one adaptive response to development program (epigenetic control).

Phenotypic plasticity occurs when the same genotype produces different phenotypes in response to varying environmental conditions (Pfennig et al., 2010; Whitman and Agrawal, 2009). Some evolutionary events are assigned to phenotypic plasticity, such as the origin of novel phenotypes, the divergence among population and species, the formation of new species and adaptive radiation (Pfennig et al., 2010). Thus, the environment could act together with genotypes and induce a change in morphology, but it could also be accompanied by a change in behaviour and physiology (Bourdeau, 2009; Cornwallis and Birkhead, 2008; Pfennig et al., 2010). Phenotypic plasticity could promote genetic accommodation and genetic assimilation. Genetic accommodation occurs when mutation or environmental changes result in novel phenotypes which result in an adaptive phenotype through quantitative genetic changes (Suzuki and Nijhout, 2006). On the other hand, when new phenotypes do not change under different environments, in this case it is genetic assimilation and the trait will be expressed constitutively over evolutionary time (Waddington, 1953). One common kind of phenotypic plasticity in butterflies is polyphenism, when individuals with identical genomes respond to different environmental cues by expressing alternatively developmental pathways (Aubin-Horth and Renn, 2009; Snell-Rood et al., 2010), which results in distinctively different adaptive phenotypes (Pfennig et al., 2010). The study of polyphenism in butterflies has been analysed from the point of view of variation in the seasonal wing colour pattern (e.g. Canfield and Pierce, 2010; Nijhout, 2003; Ruszczyk et al., 2004; Shapiro, 1976), but how wing shapes respond to polyphenism variation is still a poor-studied issue.

Another informative way to understand the genetic and environmental factors acting on wing shapes is analysing fluctuating asymmetry (FA) (Møller and Swaddle, 1997; Palmer and Strobeck, 1986). Knowing that both sides of the body of an individual share the same genome and nearly the same environment, FA results from small random perturbations in developmental processes that take place on the left and right sides of the body (Klingenberg and Nijhout, 1999). If the development of two traits is linked, the same perturbations can have an effect on both of them simultaneously and, hence, can generate a correlation between their asymmetries.

4. BIOTIC FILTERS

Direct and indirect interactions between individuals are an important component of biological systems that connect different species in the ecosystems (Jordano, 2016). Biotic filters are considered the results of ecological interactions between organisms and are divided into interactions between individuals of the same species (intraspecific) or between individual from different species (interspecific) (Fig. 1).

4.1 Intraspecific Interaction

Sexual selection has a relevant influence on the evolution of wing shapes, mainly when wings interact with reproductive behaviour and sexual communication. Due to the reproductive roles of each sex, the divergence in flight and wing shape can be found between females and males. While males spend more time obtaining mates (e.g. defending territories), females typically focus on the search for suitable foraging and oviposition sites. Additionally, females show a typical flight, being slower than males which are more mobile and has a stronger flights for escape from predators (Joron, 2005; Jones et al., 2013). Consequently, the difference in behaviours can reflect on morphological differences because males and females interact differently with biotic and abiotic factors.

Therefore, Chazot et al. (2016) suggest that the existence of strong sexual selection could contribute to wing size and shape variation among sexes. An interesting discovery is that changes in forewing shape were associated with dispersal ability only in females of the Nearctic butterfly *Melitaea cinxia* (Breuker et al., 2007). On the other hand, the aspect ratio and wing centroid measurement showed to be sex-specific in tropical forest *Morpho* butterflies (DeVries et al., 2010). For this group of butterflies, the wing shape was shown to depend on body size only in females (DeVries et al., 2010), and the aspect ratio proxy is lower in females than males of Nearctic *P. aegeria* (Berwaerts et al., 2002). In contrast, at the community level, Graça et al. (2017b) evaluating the correlation between wing–thorax ratio and forest stratum between sex in a tropical rainforest founded that was stronger in females than in males, suggesting that females a stronger relationship between phenotypic and habitat conditions than males. Moreover, the sexual dimorphism is more accentuated in understory species (Graça et al., 2017b). Taking into account the research on how the colour pattern covaries with both wing size and shape, Chazot et al. (2016)

found that forewing shape dimorphism between sexes is higher in species that show strong colour dimorphism, and flight behaviour may be selected jointly with cryptic colouration, thereby influencing both wing shape and wing colour evolution. These studies could will shed light on the extent to which different dimensions of butterfly wings respond independently to different selective pressures or are instead constrained by functional or developmental components. In addition, when females visit different microhabitats, those males and these places have Müllerian mimetic rings; there may be a differential selective pressure on females that converge for Batesian mimicry, generating sex-limited mimicry (Fig. 2). In this sense, sex-limited mimicry is more related to natural selection than sexual selection and depends on the spatial variation of the frequency of

Fig. 2 Phylogenetic relationships between some species of Dismorphiinae (Lepidoptera: Pieridae), showing a small sample of the wing shape and colour variation associated with occurrence in different microhabitats and Batesian mimicry. Note that sexual dimorphism can occur in both the wing shape and colour, including sex-limited mimicry in *Dismorphia melia* (female is a mimic of the *Actinote* species in Southeast Brazil). Phylogenetic relationships based on Wahlberg et al. (2014).

aposematic models (Kunte, 2008). This phenomenon seems to be favoured in species that already have sexual dimorphism (Timmermans et al., 2017).

4.2 Interspecific Interaction

One interesting system that can be used to evaluate wing shape is the mimicry ring. The convergence between different species for one wing pattern can be the result of a few factors, including toxicity, superficial similarity and cooccurrence with other aposematic species. In addition to the great diversity in colour and size, convergence of the wing shape at the community level is one very interesting opportunity to evaluate mimicry (Johansson et al., 2009), which can make it possible to understand these factors separately through the resemblance of mimicry rings. Recent studies have unravelled the mechanisms of structural mimetic ring communities, revealing important roles in niche differentiation and speciation processes (see Aubier et al., 2017; Joshi et al., 2017).

In Müllerian rings, comimic butterfly species share chemical defence traits which are avoided by predators, imposing strong selective pressure and thus leading to convergence onto a shared warning colour signal (Müller, 1878, 1879). It is important to note that while Müllerian mimicry is a mutualistic interaction, since all members of the ring may have benefits for survival, Batesian mimicry by palatable species is an antagonistic interaction (but see Rowland et al., 2010). Generally, the older or most abundant species in a mimicry ring drives the evolution of phenotypic resemblance in other species (Franks and Sherratt, 2007; Mallet, 1999; Ruxton et al., 2008; Turner, 1977). Thus, the warning signal in diverse communities is strongly frequency-dependent on comimetics (Chouteau et al., 2016; Joshi et al., 2017). Additionally, mimicry benefits are associated with a given warning signal, depending on the relative numbers of prey sampled compared to the available ones, and are usually driven by the most toxic or the most abundant prey species (Mallet and Joron, 1999). The ability of predators to recognize the signals of toxic prey is an important determinant of selection on resemblance in a mimicry system (Ihalainen et al., 2012; Rowe et al., 2004). Ihalainen et al. (2008) found that phenotypic similarity is influenced not only by natural selection favouring accurate mimicry but also by the genetic architecture underlying variation in phenotypes, increasing predation against imperfect mimics.

One trade-off of selective pressure on imperfect mimicry is the ability of the butterflies' escape from the predation (Fig. 1B). Butterflies with longer and larger area wings flew more slowly and were captured more easily by predators when chased (Svensson and Friberg, 2007), on the other hand, the butterflies with shorter and smaller wings tended to fly faster and more erratically and frequently escaped from predators. Graça et al. (2017a) found that tropical butterflies and birds are similar when the body size (functional component) variation is analysed, but this relationship not is detected when looking at taxonomic butterfly species composition. Bates (1862) suggested that mimicry in insects should extend to their physical motions, as should the convergence of the locomotory behaviour as well as colour patterning. Studies that compared the type of movement between two comimetic species and two other phylogenetically closely related species found that the comimetics *Heliconius cydno* and *Heliconius sapho* beat their wings more slowly and with more symmetrical motion than their respective sister species, the comimetics *Heliconius melpomene* and *H. erato* (Srygley and Ellington, 1999). Thus, wing motion appears to be one important component of the overall mimetic signal and may serve to identify mimetics before the predator can see details of the wing pattern (Srygley, 2007). This fact made wing shape variation an important factor for understanding the mimetic ring in addition to the other factors that act on this structure. Additionally, spatial cooccurrence can reinforce the phenotypic resemblance in mimicry rings (Joron and Iwasa, 2005).

Some structures that give extraordinary shapes to lepidopteran wings are related to escape strategies, such as the long tails present in different lineages of butterfly and moths (see Barber et al., 2015; López-Palafox and Cordero, 2017; Robbins, 1981; Zhong et al., 2016). The escape may also be related to mimicry, although little studied, escape or invasive mimicry has been increasingly considered as a real possibility in butterflies (see Pinheiro and Freitas, 2014; Pinheiro et al., 2016). In these mimetic systems, the fastest, best escapes are "models" for species with similar (Müllerian) or less (Batesian) effective escaping tactics (Pinheiro and Freitas, 2014). Like aposematic based mimicry, convergence in escape mimicry occurs in the wing colour patterns, but considering the key aspect of flight in this kind of mimicry, a strong convergence in wing shape or others flight-related traits would be expected. Moreover, Pinheiro et al. (2016) suggest that some mimicry rings can contain species combining the two kinds of defence and signs for predators, unpalatability and escape ability. A comparative analysis of wing shape can be a promising approach to disentangle the selective pressures in these complex mimetic systems.

5. ABIOTIC FILTERS

The emergence of wing shape diversity throughout evolution is a result of a different selective processes, such as biotic interactions and stochastic processes, but also by environmental pressure (Fig. 1). The species are locally grouped taking into consideration their traits to the environmental conditions (Cavender-Bares et al., 2009; Ricklefs, 2004), and the functional traits from species must be compatible with the habitat (Arnan et al., 2013; Graça et al., 2015). Therefore, the wing shape is partly moulded by the affinity of an organism, population or community in respect to the habitat conditions.

Wing functionality associated with wing shape that can be strongly affected by environment conditions is the capacity of the flight. Changes in wing shape allow for different aerodynamics of flight and therefore, an occurrence of diverse flight types. Considering species of bats, birds and butterflies that forage near flat and smooth surfaces, they all have modified body aerodynamics to glide in ground effect and enhance flight performance at lower costs (Cespedes et al., 2015; Kim et al., 2014). Usually, the elongate wing shape (high values of aspect ratio), mainly in the forewing, is more related with gliding flight. While the rounded overall wing shape (lower values for aspect ratio) predicted to be more manoeuvrable (Betts and Wootton, 1988) and employ flapping flight to a greater degree (Dudley, 2000).

Additionally, the wings of migratory individuals frequently have a more elongate forewing apex, which allows for a more energetic and aerodynamic flight (DeVries et al., 2010; Johansson et al., 2009; Johnson, 1969). Individual migrants, usually across very diverse places, have some special wing characteristics for flight. In *Danaus plexippus*, a population that shows migratory behaviour has larger and more elongated forewings and a higher aspect ratio than resident individuals (Satterfield and Davis, 2014). Like birds (Egbert and Belthoff, 2003), the wing shape of migrations differs considerably among dragonflies species (Johansson et al., 2009). Bird studies demonstrated that migrating species tend to have longer and more pointed wings (Kaboli et al., 2007; Mönkkönen, 1995; Voelker, 2001). Another interesting observation is that the effects of migration and mate guarding on wing shape appear to evolve (partly) independently (Johansson et al., 2009).

Long-term studies in tropical rainforest sites show that butterfly communities present vertical stratification, including the iconic *Morpho* butterflies

(DeVries and Walla, 2001; DeVries et al., 1997). DeVries et al. (2010) found that spatial use of forest microhabitats differs among *Morpho* species and that natural selection has acted to influence flight behaviours and wing morphology relative to the costs of flights. The difference in wing shape on stratification has only been observed in males (DeVries et al., 2010), suggesting that sexual dimorphism may modulate such stratification patterns (Graça et al., 2017b). The difference in sunlight level and host plant vertical distribution (Beccaloni, 1997) can be explained by understory/canopy segregation (DeVries, 1988). The evolution of wing shape has been driven by microhabitat use (Chazot et al., 2016), and flight morphology explains in part the vertical stratification of tropical fruit-feeding Nymphalidae (Graça et al., 2017b). According to Papageorgis (1975), wing colour patterns play a key role in defence, but also have to be in accordance with habitat use and the ability to regulate body temperature to prevent collapse. The relation between colour and wing shape from the perspective of thermoregulation is a fundamental subject but still little studied.

Although size variation in herbivorous species related with host plants can be attributed directly to nutritional quality (Nylin and Gotthard, 1998), this relationship does not work with shape variation. However, this shape variation is involved in static allometry (Gould, 1966), and the shape differences could be attributed to size differences resulting from starvation (Jorge et al., 2011). An extreme case of abiotic filter affecting wing shape is the evolution of brachyptery and flightlessness. Although it is a recurring phenomenon in moths, it is recorded for only one genus of Andean butterflies (Pyrcz et al., 2017; Viloria et al., 2003). This brachypterous butterfly evolved in Páramo vegetation, a high mountain habitat which presents extreme weather conditions, with strong winds and low daily temperatures, making flight dangerous and expensive. Interestingly, the reduction of the wings occurs only in females, indicating that there are constraints that keep the males with functional wings (Viloria et al., 2003).

6. MACROEVOLUTIONARY PROCESSES

All these previous selective pressures (ontogenetic, biotic and abiotic filters) acting repeatedly over a large amount of time must result in different evolutionary patterns in both species and traits. Distinct selective pressures acting on microevolutionary processes repeatedly result in different evolutionary scenarios, which take into account the occurrence of selective or nonselective pressures. The habitat can act on the trait, from sudden (natural catastrophic event), to regular (seasonal variation), or progressive events (global warming or urbanization), changing the history of wing evolution (Hill et al., 2002; Vitousek et al., 1997). In butterflies, this effect can be easily seen through the close relationship of these organisms with the environment.

Identifying exactly which factors or selective pressure acted on wing is paramount to define primary wing features, but it is not easy, due to the wide functionality attributed to this structure, since the species traits and occurrence patterns result from a series of ecological and evolutionary processes (Ackerly, 2003; Thomas et al., 2016). Nevertheless, one way to clarify how these selective pressure act on distinct aspects of the wing is observing the different facet of the wings which are related to some dimension of the fitness (growth, reproduction and survival) and access the allometric and assimetric relationships and the modularity between these functional components of wings. To obtain just the effect of these selective factors is important to remove the shared evolutionary history between the species, allowing to get the trait value without the effect of the phylogenetic relationship (Webb, 2000). One way to access ecologically functional traits is using morphological characteristics, which are usually good predictors of functional diversity in winged animals. Understanding to what extent wings evolved under different selective pressures is important to understand which factor could affect wing evolution. The capacity to fly is one of the most important functions of wings, and by using phylogenetic comparative analysis, it is possible to infer mechanisms underlying trait evolution. Therefore, we can infer the evolutionary history from the traits using some models with different evolutionary scenarios in order to understand how time, the mode and the phylogeny molded trait evolution. To test for this, these models must take into account phylogeny relationships among species and/or selective pressure on the traits.

An interesting model that can be used to simulate the evolution of the traits that consider phylogeny, but not selective pressure, was the Brownian model (BM) (Felsenstein, 1985) and the lambda model (Pagel, 1999). In the BM, change in the trait values occurred at a constant rate, without directional pressure, like genetic drift. In this view, the lambda model was frequently used for the inference of phylogenetic signals, so it is used to evaluate whether closely related species showed similar trait values. On the other hand, some models take into account selective pressures scenarios. If the attributes evolved as a result of the emergence of some barrier, for example, then the occurrence of stabilizing pressure and the attributes of each side of the barrier evolved to optimal values was considered. In this case, the evolutionary model of the attribute was the Ornstein–Uhlenbeck (OU) (Butler and King, 2004). However, when the change in the traits is exponential in terminal or ancestor branches the best fit will be the early burst (EB) (Harmon et al., 2010). Moreover, the change in the trait can be more strongly related to the speciation process, which can be tested using a kappa model (Pagel, 1999), and, finally, the null model (white noise) to compare with other models. To understand how the traits evolved on the phylogenetic context allows us to know which evolutionary processes can act on the evolution of ancient lineages and then consider the recent processes that can act.

The estimation of the phylogenetic constraint in morphological traits and in trait–habitat relationships helps us to understand the mechanisms underlying evolutionary selection and its impact on current ecological patterns (Ackerly, 2009). Some processes, like adaptive radiation, suggest that diversification is due to changes in species traits in response to ecological opportunities promoted by available niche space, while niche conservatism describes the tendency of closely related species to retain similar characteristics, minimizing evolutionary change (Ackerly, 2003, 2009; Harvey and Pagel, 1991; Losos, 2010). In part, niche conservatism reflects phylogenetic signal; however, phylogenetic signal alone does not directly denote niche conservation.

Studies using the comparative phylogenetic methods have increased in winged animals, for example, in birds (Claramunt et al., 2012), in bats (Santana and Cheung, 2016) and in butterflies. Chazot et al. (2016) evaluated the process that drove the diversification of wing size and shape in *Morpho* using the macroevolutionary perspective, but not using these evolutionary models. Furthermore, wing shape is one of the most interesting forms for understanding the differences in flight ability occurrence, and the aspect ratio is one way to obtain the wing shape for butterflies (Betts and Wootton, 1988; Breuker et al., 2007). More research, however, is needed to be performed to clarify our understanding of the evolution of the traits in butterflies.

7. CONCLUDING REMARKS

In this brief review, we have shown how an intricate eco-evolutionary history can explain the amazing morphological patterns displayed in butterfly wings (Fig. 1) by discussing some recent advances in this subject. As we have seen, the diversity of wing shape and colour patterns could evolve from different selective pressures. These pressures begin in the early stages, when biotic and abiotic filters constrain the ontogenetic process which will reflect on structures of adult butterflies, including wing morphology. After becoming adults, butterflies have no opportunity to change their exoskeleton; however, other selective forces can act at this stage, such as sexual selection, when adaptive shapes could to be selected for sexual partners, or through the natural selection exerted by predators where some genes associated with wing shape could to be removed or remain in the population over time. Apart from these pressures, environmental conditions impose a strong abiotic filter on aerodynamic traits, including wing shape. In summary, the interaction between genotype, phenotype and environment could answer many questions about the evolutionary history of wing shapes. Obviously, these features are not restricted to butterflies, but we propose that they can be more easily understood by using these organisms as models.

Nevertheless, disentangling which factors are operating in these different contexts is the a major challenge. The use of the phylogenetic approach is fundamental to tease apart ecological factors operating in wing shapes from shared evolutionary history. Moreover, some morphological tools allow for understanding the asymmetric, allometric and modularity effects that act on the wing shapes, while separating each effect. Allometry allows for the comparison of growth differentials between different parts of the body in males and females, in different stages in ontogenies, or different structures (modularity). In the mimicry context, for example, the allometric analysis shows that conspicuous aposematic red spots and wing shapes grow in a similar way, but sex-related differences can occur between species in the *Heliconius* postman ring Müllerian mimicry (Rossato et al., in press). This kind of approach can be one important way of understanding the selective pressures acting in both intra- and interspecific communications.

The convergence in mimicry rings has a clear spatial structure, consequently, the spatial distribution of mimetic rings and predators can help us to understand the role of space in wing convergence. The use of GM as a morphoecological tool to study wing shape is quite recent, especially in the context of mimicry evolution (Jones et al., 2013; Mérot et al., 2016; Rossato et al., in press). A good way to begin a wing shape study is to choose an informative aspect of wing functionality (e.g. reproduction, dispersion, thermoregulation, communication) to be investigated in a mimicry ring and look for others selective pressures that can act on it and include it in the analysis, and trying to separate the factors and see how they affect the trait. Thus, if an investigator chooses to evaluate wing shape regarding spatial phenotypic convergence, it is not only important to compare similarities between species, but the flight capacity to escape the predator must be taken into account, since this selective pressure acts together with convergence pressure. Therefore, it can be easily understood how these factors work: the way this will be selected in the wing will be a trade-off between the importance of a certain feature to perform the communication function or to the flight. In the end, the wing will be the result of these two, or more, pressures acting on the shape of the wing.

Although butterflies are models in many areas of evolutionary biology, most studies are restricted to a few well-studied lineages, such as the large swallowtail butterflies (Papilionidae), or the well-known *Heliconius*. In fact, these model groups are only a small fraction of the great radiation of butterflies that have occurred since the Cretaceous (Heikkilä et al., 2012). When we speak of understanding the wing shape diversity in butterflies, we would like to include all of this variation. In this sense, we need to know more about some neglected lineages, to unravel the ecological and evolutionary patterns that operate in the less studied families. In recent years, robust phylogenetic hypotheses are being obtained for almost all butterfly families, which can serve as the basis for these evolutionary studies (e.g. Sahoo et al., 2016; Seraphim et al., 2018; Wahlberg et al., 2014). Identifying new model systems is also crucial, for example, the Dismorphiinae that illuminated early ideas about mimicry (Bates, 1862) are a candidate group to study the evolution of wing shapes (Fig. 2). This group includes species with slight sexual dimorphism in wing shape, such as the *Pseudopieris* species, which present a typical pierid bauplan and occur in more naturally disturbed open edge environments. Also, it includes species with strong sexual dimorphism, such as *Enantia* species, whose males present unusually short fore wings. And finally, it includes *Dismorphia* and other genera which are Batesian mimics (Bates, 1862; LeCrom et al., 2004), including species with both sex-limited dimorphism (males flying in distinct microhabitats from females) or without dimorphism in the colour pattern (males and females flying in same microhabitats), but with marked dimorphism in the wing shape due to male investment in androconial scales.

In addition to these evolutionary studies with closely related lineages, there are major questions about the wings of butterflies. For example, what are the great convergences and divergences imposed by different biotic (e.g. host plants, interaction with ants, predation rates) and abiotic filters (e.g. insularity, humidity, altitude, etc.) that explain this diversity of wing shape in butterfly families? How is this shape diversity restricted by the evolutionary history of the groups, including phylogenetic constraints and biogeography? The answers to these considerable issues are unlikely to be obtained from specific studies on models, but can be revealed by comparative studies on a large scale.

The emergence of next-generation sequencing can allow us to find specific parts of the genome that can be related to wing shape, though, for example, the quantitative trait locus (Lynch and Walsh, 1998). Another promising approach is evaluating the epigenetic factors that can act on the genome, resulting in wing shape changes in different environments (Jablonka, 2017). The possibility of integrative studies joining genomics, bioinformatics, field experiments and mathematical modelling seems almost infinite and points to a promising future in all areas of evolutionary biology, including ecomorphology and ecophysiology. Almost two centuries after Bates's insights about the biological relevance of studying the butterfly wings, we can verify that he was right, and that finally scientists are close to solving many mysteries that are still waiting to be unveiled in the flapping wings of butterflies.

ACKNOWLEDGEMENTS

We would like to thank Richard ffrench-Constant for the invitation to write about some interesting issue. D.O.R. and L.A.K. are indebted to Gilson R.P. Moreira for his introduction to study insect morphology. We are also grateful to Derrick Mulder for the edit the text. D.O.R. was supported by a CAPES PhD Program Fellowship. L.A.K. was supported by PNPD/CAPES Program Fellowship. L.D. research activities have been supported by CNPq Productivity Fellowship (Grant 307886/2015-8). L.D. research has been developed in the context of National Institutes for Science and Technology (INCT) in Ecology, Evolution and Biodiversity Conservation, supported by MCTIC/CNPq (proc. 465610/2014-5) and FAPEG.

REFERENCES

Ackerly, D.D., 2003. Community assembly, niche conservatism and adaptive evolution in changing environments. Int. J. Plant Sci. 164, S165–S184.

Ackerly, D.D., 2009. Conservatism and diversification of plant functional traits: evolutionary rates versus phylogenetic signal. PNAS 106, 19699–19706.

Adams, D.C., Rohlf, F.J., Slice, D.E., 2004. Geometric morphometrics: ten years of progress following the "revolution". Ital. J. Zool. 71, 5–16.

Alexander, D.E., 2015. On the Wing: Insects, Pterosaurs, Birds, Bats and the Evolution of Animal Flight. Oxford University Press, Oxford.

Allen, C.E., Beldade, P., Zwaan, B.J., Brakefield, P.M., 2008. Differences in the selection response of serially repeated color pattern characters: standing variation, development, and evolution. BMC Evol. Biol. 8, 94.

Arnan, X., Cerdá, X., Rodrigo, A., Retana, J., 2013. Response of ant functional composition to fire. Ecography 36, 1182–1192.
Aubier, T.G., Elias, M., Llaurens, V., Chazot, N., 2017. Mutualistic mimicry enhances species diversification through spatial segregation and extension of the ecological niche space. Evolution 4, 826–844.
Aubin-Horth, N., Renn, S.C.P., 2009. Genomic reaction norms: using integrative biology to understand molecular mechanisms of phenotypic plasticity. Mol. Ecol. 18, 3763–3780.
Baena-Lopez, L.A., Garcia-Bellido, A., 2006. Control of growth and positional information by the graded vestigial expression pattern in the wing of *Drosophila melanogaster.* PNAS 103, 13734–13739.
Bai, M., Beutel, R.G., Song, K.Q., Liu, W.G., Malqin, H., Li, S., Hu, X.Y., Yang, X.K., 2012. Evolutionary patterns of hind wing morphology in dung beetles (Coleoptera: Scarabaeinae). Arthropod Struct. Dev. 41, 505–513.
Baker, R.R., 1972. Territorial behaviour of the nymphalid butterflies, *Aglais urticae* (L.) and. *Inachis io* (L.). J. Anim. Ecol. 41, 453–469.
Barber, J.R., Leavell, B.C., Keener, A.L., Breinholt, J.W., Chadwell, B.A., McClure, C.J.W., Hill, G.M., Kawahara, A.Y., 2015. Moth tails divert bat attack: evolution of acoustic deflection. PNAS 112, 2812–2816.
Bates, H.W., 1862. Contributions to an insect fauna of the Amazon valley (Lepidoptera: Heliconidae). Trans. Linn. Soc. Lond. 23, 495–566.
Bates, H.W., 1864. The Naturalist on the River Amazons. A Record of Adventures, Habits of Animals, Sketches of Brazilian and Indian Life, and Aspects of Nature Under the Equator, During Eleven Years of Travel. J. Murray, London.
Beccaloni, G.W., 1997. Vertical stratification of ithomiine butterfly (Nymphalidae: Ithomiinae) mimicry complexes: the relationship between adult flight height and larval host-plant height. Biol. J. Linn. Soc. 62, 313–341.
Beldade, P., Brakefield, P.M., 2002. The genetics and evo-devo of butterfly wing patterns. Nat. Rev. Genet. 3, 442–452.
Berwaerts, K., Van Dyck, H., Aerts, P., 2002. Does flight morphology relate to flight performance? An experimental test with the butterfly *Pararge aegeria.* Funct. Ecol. 16, 484–491.
Betts, C., Wootton, R., 1988. Wing shape and flight behaviour in butterflies (Lepidoptera: Papilionoidea and Hesperioidea): a preliminary analysis. J. Exp. Biol. 138, 271–288.
Blankers, T., Adams, D.C., Wiens, J.J., 2012. Ecological radiation with limited morphological diversification in salamanders. J. Evol. Biol. 25, 634–646.
Bonebrake, T.C., Ponisio, L.C., Boggs, C.L., Ehrlich, P.R., 2010. More than just indicators: a review of tropical butterfly ecology and conservation. Biol. Conserv. 143, 1831–1841.
Bookstein, F.L., 1991. Morphometric Tools for Landmark Data Geometry and Biology. Cambridge University Press, Cambridge.
Bots, J., Breuker, C.J., Kaunisto, K.M., Koskimäki, J., Gossum, H.V., Suhonen, J., 2012. Wing shape and its influence on the outcome of territorial contests in the damselfly *Calopteryx virgo.* J. Insect Sci. 12, 96.
Bourdeau, P.E., 2009. Prioritized phenotypic responses to combined predators in a marine snail. Ecology 90, 1659–1669.
Breuker, C.J., Debat, V., Klingenberg, C.P., 2006. Functional evo-devo. TREE 21, 488–492.
Breuker, C., Brakefield, P.M., Gibbs, M., 2007. The associations between wing morphology and dispersal is sex-limited in the *Glanville* fritillary butterfly. Eur. J. Entomol. 104, 445–452.
Breuker, C.J., Gibbs, M., van Dongen, S., Merckx, T., Van Dyck, H., 2010. The use of geometric morphometrics in studying butterfly wings in an evolutionary ecological context. In: Elewa, A.M.T. (Ed.), Morphometrics for Nonmorphometricians. Springer-Verlag, Berlin, pp. 271–287.

Brown Jr., K.S., 1997. Diversity, disturbance, and sustainable use of Neotropical forests: insects as indicators for conservation monitoring. J. Insect. Cons. 1, 25–42.
Butler, M.A., King, A.A., 2004. Phylogenetic comparative analysis: a modeling approach for adaptive evolution. Am. Nat. 164, 683–695.
Cadotte, M.W., Davies, J., 2016. Phylogenies in Ecology: A Guide to Concepts and Methods. Princeton University Press, Princeton and Oxford.
Canfield, M.R., Pierce, N.E., 2010. Facultative mimicry? The evolutionary significance of seasonal forms in several Indo-Australian butterflies in the family Pieridae. Trop. Lepid. 20, 1–7.
Carreira, V.P., Soto, I.M., Mensch, J., Fanara, J.J., 2011. Genetic basis of wing morphogenesis in *Drosophila*: sexual dimorphism and non-allometric effects of shape variation. BMC Dev. Biol. 11, 32.
Cavender-Bares, J., Kozak, K.H., Fine, P.V.A., Kembel, S.W., 2009. The merging of community ecology and phylogenetic biology. Ecol. Let. 12, 693–715.
Cespedes, A., Penz, C.M., DeVries, P.J., 2015. Cruising the rain forest floor: butterfly wing shape evolution and gliding in ground effect. J. Anim. Ecol. 84, 808–816.
Chazot, N., Panara, S., Zilbermann, N., Blandin, P., Le Poul, Y., Cornette, R., Elias, M., Debat, V., 2016. Morpho morphometrics: shared ancestry and selection drive the evolution of wing size and shape in *Morpho* butterflies. Evolution 70, 181–194.
Cheverud, J.M., 1996. Developmental integration and the evolution of pleiotropy. Am. Zool. 36, 44–50.
Chouteau, M., Arias, M., Joron, M., 2016. Warning signals are under positive frequency-dependent selection in nature. PNAS 113, 2164–2169.
Claramunt, S., Derryberry, E.P., Brumfield, R.T., Remsen, J.V., 2012. Ecological opportunity and diversification in a continental radiation of birds: climbing adaptations and cladogenesis in the Furnariidae. Am. Nat. 179, 649–666.
Cock, A.G., 1966. Genetical aspects of metrical growth and form in animals. Q. Rev. Biol. 41, 131–190.
Cornwallis, C.K., Birkhead, T.R., 2008. Plasticity in reproductive phenotypes reveals status-specific correlations between behavioral, morphological, and physiological sexual traits. Evolution 62, 1149–1161.
Davies, N.B., 1978. Territorial defense in the speckled wood butterfly (*Pararge aegeria*): the resident always wins. Ani. Behav. 26, 138–147.
De Keyser, R., Breuker, C.J., Hails, R.S., Dennis, R.L.H., Shreeve, T.G., 2015. Why small is beautiful: wing colour is free from thermoregulatory constraint in the small lycaenid butterfly, *Polyommatus icarus*. PLoS One 10, e0122623.
DeVries, P.J., 1988. Stratification of fruit-feeding nymphalid butterflies in a Costa Rican rainforest. J. Res. Lep. 26, 98–108.
DeVries, P.J., Walla, T.R., 2001. Species diversity and community structure in neotropical fruit-feeding butterflies. Biol. J. Linn. Soc. 74, 1–15.
DeVries, P.J., Murray, D., Lande, R., 1997. Species diversity in vertical, horizontal, and temporal dimensions of a fruit-feeding butterfly community in an Ecuadorian rainforest. Biol. J. Linn. Soc. 62, 343–364.
DeVries, P.J., Penz, C.M., Hill, R.I., 2010. Vertical distribution, flight behaviour and evolution of wing morphology in *Morpho* butterflies. J. Anim. Ecol. 79, 1077–1085.
Dincă, V., Dapporto, L., Vila, R., 2011. A combined genetic-morphometric analysis unravels the complex biogeographical history of *Polyommatus icarus* and *Polyommatus celina* common blue butterflies. Mol. Ecol. 20, 3921–3935.
Dudley, R., 2000. The Biomechanics of Insect Flight: Form, Function and Evolution. Princeton University Press, Princeton.
Egbert, J.R., Belthoff, J.R., 2003. Wing shape in house finches differs relative to migratory habit in eastern and western North America. Condor 105, 825–829.

Felsenstein, J., 1985. Phylogenies and the comparative method. Am. Nat. 125, 1–15.
Ferson, S., Rohlf, F.J., Koehn, R.K., 1985. Measuring shape variation of two-dimensional outlines. Syst. Zool. 34, 59–68.
Frankino, W.A., Zwaan, B.J., Stern, D.L., Brakefield, P.M., 2007. Internal and external constraints in the evolution of morphological allometries in a butterfly. Evolution 61, 2958–2970.
Franks, D.W., Sherratt, T.N., 2007. The evolution of multicomponent mimicry. J. Theor. Biol. 244, 631–639.
Freitas, A.V.L., Marini-Filho, O.J., 2011. Plano de Ação Nacional para a Conservação dos Lepidópteros Ameaçados de Extinção. Instituto Chico Mendes de Conservação da Biodiversidade (ICMBio), Brasília.
Gillies, T.E., Cabernard, C., 2011. Cell division orientation in animals. Curr. Biol. 21, R599–R609.
Gould, S.J., 1966. Allometry and size in ontogeny and phylogeny. Biol. Rev. 41, 587–638.
Graça, M.B., Morais, J.W., Franklin, E., Pequeno, P.A.C.L., Souza, J.L.P., Bueno, A.S., 2015. Combining taxonomic and functional approaches to unravel the spatial distribution of an Amazonian butterfly community. Environ. Entomol. 45, 301–309.
Graça, M.B., Pequeno, P.A.C.L., Franklin, E., Souza, J.L.P., Morais, J.W., 2017a. Taxonomic, functional, and phylogenetic perspectives on butterfly spatial assembly in northern Amazonia. Ecol. Entomol. 42, 816–826. https://doi.org/10.1111/een.12454.
Graça, M.B., Pequeno, P.A.C.L., Franklin, E., Morais, J.W., 2017b. Coevolution between flight morphology, vertical stratification and sexual dimorphism: what can we learn from tropical butterflies? J. Evol. Biol. 30, 1862–1871.
Greenslade, P., Farrow, R.A., Smith, J.M.B., 1999. Long distance migration of insects to a subantarctic island. J. Biogeogr. 26, 1161–1167.
Grimaldi, D., Engel, M.S., 2005. Evolution of the Insects. Cambridge University Press, Cambridge, New York, Melbourne.
Grodnitsky, D.L., Dudley, R., Gilbert, L., 1994. Wing decoupling in hovering flight of swallowtail butterflies (Lepidoptera: Papilionidae). Trop. Lepid. 5, 85–86.
Harmon, L.J., Losos, J.B., Davies, T.J., Gillespie, R.G., Gittleman, J.L., Jennings, W.B., Kozak, K.H., McPeek, M.A., Moreno-Roark, F., Near, T.J., Purvis, A., Ricklefs, R.E., Schluter, D., Schulte, J.A., Seehausen, O., Sidlauskas, B.L., Torres-Carvajal, O., Weir, J.T., Mooers, A., 2010. Early bursts of body size and shape evolution are rare in comparative data. Evolution 64, 2385–2396.
Harvey, P.H., Pagel, M.D., 1991. The Comparative Method in Evolutionary Biology. Oxford University Press, New York.
Hassall, C., 2015. Strong geographical variation in wing aspect ratio of a damselfly, *Calopteryx maculata* (Odonata: Zygoptera). PeerJ 3, e1219. https://doi.org/10.7717/peerj.1219.
Heikkilä, M., Kaila, L., Mutanen, M., Peña, C., Wahlberg, N., 2012. Cretaceous origin and repeated tertiary diversification of the redefined butterflies. Proc. R. Soc. B 279, 1093–1099.
Hill, J.K., Thomas, C.D., Fox, R., Telfer, M.G., Willis, S.G., Asher, J., Huntley, B., 2002. Responses of butterflies to twentieth century climate warming: implications for future ranges. Proc. R. Soc. B 269, 2163–2171.
Holwell, G.I., Herberstein, M.E., 2010. Chirally dimorphic male genitalia in praying mantids (Ciulfina: Liturgusidae). J. Morphol. 271, 1176–1184.
Houle, D., Mezey, J., Galpern, P., Carter, A., 2003. Automated measurement of *Drosophila* wings. BMC Evol. Biol. 3, 25.
Huang, S.S., Huang, J.S., 2005. TGF-beta control of cell proliferation. J. Cell. Biochem. 96, 447–462.
Huxley, J., Teissier, G., 1936. Terminology of relative growth rates. Nature 137, 780–781.

Ihalainen, E., Lindstroöm, L., Mappes, J., Puolakkainen, S., 2008. Butterfly effects in mimicry? Combining signal and taste can twist the relationship of Müllerian co-mimics. Behav. Ecol. Sociobiol. 62, 1267–1276.
Ihalainen, E., Rowland, H.M., Speed, M.P., Ruxton, G.D., Mappes, J., 2012. Prey community structure affects how predators select for Müllerian mimicry. Proc. R. Soc. B 279, 2099–2105.
Jablonka, E., 2017. The evolutionary implications of epigenetic inheritance. Interface Focus 7, 20160135.
Jantzen, B., Eisner, T., 2008. Hindwings are unnecessary for flight but essential for execution of normal evasive flight in Lepidoptera. PNAS 105, 16636–16640.
Johansson, F., Soderquist, M., Bokma, F., 2009. Insect wing shape evolution: independent effects of migratory and mate guarding flight on dragonfly wings. Biol. J. Linn. Soc. 97, 362–372.
Johnson, C.G., 1969. Migration and Dispersal of Insects by Flight. Meuthen, London.
Jones, R.T., Salazar, P.A., ffrench-Constant, R.H., Jiggins, C.D., Joron, M., 2012. Evolution of a mimicry supergene from a multilocus architecture. Proc. R. Soc. B 279, 316–325.
Jones, R.T., Le Poul, Y., Whibley, A.C., Mérot, C., ffrench-Constant, R.H., Joron, M., 2013. Wing shape variation associated with mimicry in butterflies. Evolution 67, 2323–2334.
Jordano, P., 2016. Chasing ecological interactions. PLoS Biol. 14, e1002559.
Jorge, L.R., Cordeiro-Estrela, P., Klaczko, L.B., Moreira, G.R.P., Freitas, A.V.L., 2011. Host-plant dependent wing phenotypic variation in *Heliconius erato*. Biol. J. Linn. Soc. 102, 765–774.
Joron, M., 2005. Polymorphic mimicry, microhabitat use, and sex-specific behaviour. J. Evol. Biol. 18, 547–556.
Joron, M., Iwasa, Y., 2005. The evolution of a Müllerian mimic in a spatially distributed community. J. Theor. Biol. 237, 87–103.
Joshi, J., Prakash, A., Kunte, K., 2017. Evolutionary assembly of communities in butterfly mimicry rings. Am. Nat. 189, E58–E76.
Kaboli, M., Aliabadian, M., Guillaumet, A., Roselaar, C.S., Prodon, R., 2007. Ecomorphology of the wheatears (genus *Oenanthe*). Ibis 149, 792–805.
Kim, E.J., Wolf, M., Ortega-Jimenez, V.M., Dudley, R., 2014. Hovering performance of Anna's hummingbirds (*Calypte anna*) in ground effect. J. R. Soc. Interface 11, 20140505.
Kingsolver, J.G., 1985. Butterfly engineering. Sci. Am. 253, 106–113.
Klingenberg, C.P., 2008. Morphological integration and developmental modularity. Annu. Rev. Ecol. Evol. Syst. 39, 115–132.
Klingenberg, C.P., 2009. Morphometric integration and modularity in configurations of landmarks: tools for evaluating a priori hypotheses. Evol. Dev. 11, 405–421.
Klingenberg, C.P., Nijhout, H.F., 1999. Genetics of fluctuating asymmetry: a developmental model of developmental instability. Evolution 53, 358–375.
Klingenberg, C.P., Zaklan, S.D., 2000. Morphological integration between development compartments in the *Drosophila* wing. Evolution 54, 1273–1285.
Klingenberg, C.P., Zimmermann, M., 1992. Static, ontogenetic and evolutionary allometry: a multivariate comparison in nine species of water striders. Am. Nat. 140, 601–620.
Klingenberg, C.P., Badyaev, A.V., Sowry, S.M., Beckwith, N.J., 2001. Inferring developmental modularity from morphological integration: analysis of individual variation and asymmetry in bumblebee wings. Am. Nat. 157, 11–23.
Klingenberg, C.P., Debat, V., Roff, D.A., 2010. Quantitative genetics of shape in cricket wings: developmental integration in a functional structure. Evolution 64, 2935–2951.
Kronforst, M.R., Papa, R., 2015. The functional basis of wing patterning in *Heliconius* butterflies: the molecules behind mimicry. Genetics 200, 1–19.

Kunte, K., 2008. Mimetic butterflies support Wallace's model of sexual dimorphism. Proc. R. Soc. B 275, 1617–1624.

LeCrom, J.F., Llorente-Bousquets, J., Constantino, L.M., Salazar, J., 2004. Mariposas de Colombia, Tomo 2: Pieridae. Carlec Ltd.; Bogotá

López-Palafox, T.G., Cordero, C.R., 2017. Two-headed butterfly vs. mantis: do false antennae matter? Peer J 5, e3493.

Losos, J.B., 2010. Adaptive radiation, ecological opportunity, and evolutionary determinism. Am. Nat. 175, 623–639.

Lynch, M., Walsh, J.B., 1998. Genetics and Analysis of Quantitative Traits. Sinauer Associates, Sunderland.

Magurran, A.E., 2004. Measuring Biological Diversity. Blackwell, Maldan.

Mallet, J., 1999. Causes and consequences of a lack of coevolution in Müllerian mimicry. Evol. Ecol. 13, 777–806.

Mallet, J., Joron, M., 1999. Evolution of diversity in warning color and mimicry: polymorphisms, shifting balance and speciation. Annu. Rev. Ecol. Syst. 30, 201–233.

Marchant, S., Moran, A.L., Marko, P.B., 2015. Out-of-the tropics or trans-tropical dispersal? The origins of the disjunct distribution of the gooseneck barnacle *Pollicipes elegans*. Front. Zool. 12, 39.

Marinari, E., Mehonic, A., Curran, S., Gale, J., Duke, T., Baum, B., 2012. Live-cell delamination counterbalances epithelial growth to limit tissue overcrowding. Nature 484, 542–545.

Mérot, C., Poul, Y.L., Therv, M., Joron, M., 2016. Refining mimicry: phenotypic variation tracks the local optimum. J. Anim. Ecol. 85, 1056–1069.

Møller, A.P., Swaddle, J.P., 1997. Asymmetry, Developmental Stability and Evolution. Oxford University Press, Oxford.

Mönkkönen, M., 1995. Do migrant birds have more pointed wings? A comparative study. Evol. Ecol. 9, 520–528.

Moreira, G.R.P., Silva, D.S., Gonçalves, G.L., 2017. Comparative morphology of the prothoracic leg in heliconian butterflies: tracing size allometry, podite fusions and losses in ontogeny and phylogeny. Arthropod Struct. Dev. 46, 462–471.

Müller, F., 1878. Uber die vortheile der mimicry bei schmetterlingen. Zool. Anz. 1, 54–55.

Müller, F., 1879. *Ituna* and *Thyridia*: a remarkable case of mimicry in butterflies. Trans. Entomol. Soc. Lond. 1879, 20–29.

Nijhout, H.F., 2003. Development and evolution of adaptive polyphenisms. Evol. Dev. 5, 9–18.

Nijhout, H.F., Grunert, L.W., 2002. Bombyxin is a growth factor for wing imaginal disks in Lepidoptera. PNAS 99, 15446–15450.

Nijhout, H.F., Smith, W.A., Schachar, I., Subramanian, S., Tobler, A., Grunert, L.W., 2007. The control of growth and differentiation of the wing imaginal disks of *Manduca sexta*. Dev. Biol. 302, 569–576.

Nijhout, H.F., Cinderella, M., Grunert, L.W., 2014. The development of wing shape in Lepidoptera: mitotic density, not orientation, is the primary determinant of shape. Evol. Dev. 16, 68–77.

Norberg, U.M., 1990. Vertebrate Flight: Mechanics, Physiology, Morphology, Ecology and Evolution. Springer, Berlin.

Nylin, S., Gotthard, K., 1998. Plasticity in life-history traits. Ann. Rev. Entomol. 43, 63–83.

O'Keefe, D., Thomas, S., Bolin, K., Griggs, E., Edgar, B., Buttitta, L., 2012. Combinatorial control of temporal gene expression in the *Drosophila* wing by enhancers and core promoters. BMC Genomics 13, 498.

Olson, E., Miller, R., 1958. Morphological Integration. University of Chicago Press, Chicago.

Outomuro, D., Johansson, F.A., 2017. Potential pitfall in studies of biological shape: does size matter? J. Anim. Ecol. 86, 1447–1457. https://doi.org/10.1111/1365-2656.12732.

Outomuro, D., Adams, D.C., Johansson, F., 2013. Wing shape allometry and aerodynamics in calopterygid damselflies: a comparative approach. BMC Evol. Biol. 13, 1–11.
Pagel, M., 1999. Inferring the historical patterns of biological evolution. Nature 401, 877–884.
Palmer, A.R., Strobeck, C., 1986. Fluctuating asymmetry: measurement, analysis, patterns. Ann. Rev. Ecol. Syst. 17, 391–421.
Papageorgis, C., 1975. Mimicry in Neotropical butterflies: why are there so many different wing-colocation complexes in one place? American Sci. 63, 522–532.
Pfennig, D.W., Wund, M.A., Snell-Rood, E.C., et al., 2010. Phenotypic plasticity's impacts on diversification and speciation. TREE 25, 459–467.
Pinheiro, C.E.G., Freitas, A.V.L., 2014. Some possible cases of escape mimicry in Neotropical butterflies. Neotrop. Entomol. 43, 393–398.
Pinheiro, C.E.G., Freitas, A.V.L., Campos, V.C., DeVries, P.J., Penz, C.M., 2016. Both palatable and unpalatable butterflies use bright colors to signal difficulty of capture to predators. Neotrop. Entomol. 45, 107–113.
Pyrcz, T.W., Lorenc-Brudecka, J., Zubek, A., Boyer, P., Gabaldon, M.C., Mavarez, J., 2017. Taxonomy, phylogeny and distribution of the genus *Steromapedaliodes* sensu novo in the Cordillera de Mérida, Venezuela (Lepidoptera: Nymphalidae: Satyrinae: Satyrini). Art. Syst. Phylo. 75, 195–243.
Regal, P.J., 1977. Ecology and evolution of flowering plant dominance. Science 196, 622–629.
Ricklefs, R.E., 2004. A comprehensive framework for global patterns in biodiversity. Ecol. Lett. 7, 1–15.
Robbins, R.K., 1981. The "false head" hypothesis: predation and wing pattern variation of Lycaenid butterflies. Am. Nat. 118, 770–775.
Rohlf, F.J., Slice, D., 1990. Extension of the Procrustes method for the optimal superimposition of landmarks. Syst. Zoo. 39, 40–59.
Rossato, D.O., Boligon, D., Fornel, R., Kronforst, M.R., Goncalves, G.L., Moreira, G.R.P., Subtle variation in size and shape of the whole forewing and the red band among comimics revealed by geometric morphometric analysis in *Heliconius* butterflies, *Ecol. Evol.* (in press).
Rowe, C., Lindström, L., Lyytinen, A., 2004. The importance of pattern similarity between Müllerian mimics in predator avoidance learning. Proc. R. Soc B271, 407–413.
Rowland, H.M., Mappes, J., Ruxton, G.D., Speed, M.P., 2010. Mimicry between unequally defended prey can be parasitic: evidence for quasi-batesian mimicry. Ecol. Lett. 13, 1494–1502.
Ruszczyk, A., Motta, P.C., Barros, R.L., Araújo, A.M., 2004. Ecological correlates of polyphenism and gregarious roosting in the grass yellow butterfly *Eurema elathea* (Pieridae). Braz. J. Biol. 64, 1–14.
Ruxton, G.D., Franks, D.W., Balogh, A.C.V., Leimar, O., 2008. Evolutionary implications of the form of predator generalization for aposematic signals and mimicry in prey. Evolution 62, 2913–2921.
Sahoo, R.K., Warren, A.D., Wahlberg, N., Brower, A.V.Z., Lukhtanov, V.A., Kodandaramaiah, U., 2016. Ten genes and two topologies: an exploration of higher relationships in skipper butterflies (Hesperiidae). PeerJ 4, e2653.
Santana, S.E., Cheung, E., 2016. Go big or go fish: morphological specializations in carnivorous bats. Proc. R. Soc. B 283, 20160615.
Satterfield, D.A., Davis, A.K., 2014. Variation in wing characteristics of monarch butterflies during migration: earlier migrants have redder and more elongated wings. Anim. Migr. 2, 1–7.
Seraphim, N., Kaminski, L.A., DeVries, P.J., Penz, C.M., Callaghan, C., Wahlberg, N., Silva-Brandão, K.L., Freitas, A.V.L., 2018. Molecular phylogeny and higher systematics of the metalmark butterflies (Lepidoptera: Riodinidae). Syst. Entomol. (in press).

Shapiro, A.M., 1976. Seasonal polyphenism. Evol. Biol. 9, 259–333.
Snell-Rood, E.C., Cash, A., Han, M.V., Kijimoto, T., Andrews, J., Moczek, A.P., 2010. Developmental decoupling of alternative phenotypes: insights from the transcriptomes of horn-polyphenic beetles. Evolution 65, 231–245.
Sobral, F.L., Cianciaruso, M.V., 2012. Estrutura filogenética e funcional de assembleias: (re) montando a ecologia de comunidades em diferentes escalas espaciais. Biosci. J. 28, 617–631.
Srygley, R.B., 1994. Locomotor mimicry in butterflies? The associations of positions of centres of mass among groups of mimetic, unprofitable prey. Philos. Trans. R. Soc. Lond. Ser. B. 343, 145–155.
Srygley, R.B., 2007. Evolution of the wave: aerodynamic and aposematic functions of butterfly wing motion. Proc. R. Soc. B 274, 913–917.
Srygley, R.B., Ellington, C.P., 1999. Discrimination of flying mimetic, passion-vine butterflies *Heliconius*. Proc. R. Soc. B 266, 2137–2140.
Suzuki, Y., Nijhout, H.F., 2006. Evolution of a polyphenism by genetic accommodation. Science 311, 650–652.
Svensson, E.I., Friberg, M., 2007. Selective predation on wing morphology in sympatric damselflies. Am. Nat. 170, 101–112.
Thomas, M.K., Kremer, C.T., Litchman, E., 2016. Environment and evolutionary history determine the global biogeography of phytoplankton temperature traits. Glob. Ecol. Biogeogr. 25, 75–86.
Timmermans, M.J., Thompson, M.J., Collins, S., Vogler, A.P., 2017. Independent evolution of sexual dimorphism and female-limited mimicry in swallowtail butterflies (*Papilio dardanus* and *Papilio phorcas*). Mol. Ecol. 26, 1273–1284.
Turner, J.R.G., 1977. Butterfly mimicry—the genetical evolution of an adaptation. Evol. Biol. 10, 163–206.
Van der Molen, S., Martínez, N.A., González-José, R., 2007. Introducción a la Morfometría Geométrica. Curso teórico práctico. Centro Nacional Patagónico and University of Barcelona, Barcelona.
Viloria, A.L., Pyrcz, T.W., Wojtusiak, J., Ferrer-Paris, J.R., Beccaloni, G.W., Sattler, K., Lees, D.C., 2003. A brachypterous butterfly? Proc. R. Soc. B 270, 21–24.
Vincent, J.P., Fletcher, A.G., Baena-Lopez, L.A., 2013. Mechanisms and mechanics of cell competition in epithelia. Nat. Rev. Mol. Cell Biol. 14, 581–591.
Violle, C., Navas, M.L., Vile, D., Kazakou, E., Fortunel, C., Hummel, I., Garnier, E., 2007. Let the concept of trait be functional!. Oikos 116, 882–892.
Vitousek, P.M., Mooney, H.A., Lubchenco, J., Melillo, J.M., 1997. Human domination of Earth's ecosystems. Science 277, 494–499.
Voelker, G., 2001. Morphological correlates of migratory distance and flight display in the avian genus *Anthus*. Biol. J. Linn. Soc. 73, 425–435.
Waddington, C.H., 1953. Genetic assimilation of an acquired character. Evolution 7, 118–126.
Wahlberg, N., Rota, J., Braby, M.F., Pierce, N.P., Wheat, C.W., 2014. Revised systematics and higher classification of pierid butterflies (Lepidoptera: Pieridae) based on molecular data. Zool. Scr 43, 641–650.
Webb, C.O., 2000. Exploring the phylogenetic structure of ecological communities: an example for rain forest trees. Am. Nat. 156, 145–155.
Weber, K.E., 1992. How small are the smallest selectable domains of form? Genetics 130, 345–353.
Whitman, D.W., Agrawal, A.A., 2009. What is phenotypic plasticity and why is it important? In: Whitman, D.W., Ananthakrishnan, T.N. (Eds.), Phenotypic Plasticity of Insects. Science Publishers, CRC Press, Enfield, pp. 1–63.
Wootton, R.J., 1992. Functional morphology of insect wings. Annu. Rev. Entomol. 37, 113–140.

Wootton, R.J., 2002. Design, function and evolution in the wings of holometabolous insects. Zool. Scr. 31, 31–40.
Zhong, M., Hill, G.M., Gomez, J.P., Plotkin, D., Barber, J.R., Kawahara, A.Y., 2016. Quantifying wing shape and size of saturniid moths with geometric morphometrics. J. Lep. Soc. 70, 99–107.

CHAPTER THREE

CRISPR/Cas9 as the Key to Unlocking the Secrets of Butterfly Wing Pattern Development and Its Evolution

Luca Livraghi*, Arnaud Martin†, Melanie Gibbs‡, Nora Braak*, Saad Arif*, Casper J. Breuker*
*Oxford Brookes University, Oxford, United Kingdom
†The George Washington University, Washington, DC, United States
‡NERC Centre for Ecology & Hydrology, Wallingford, United Kingdom

Contents

Abstract

With the exception of a few moth and butterfly species, gene-editing tools in Lepidoptera have been lagging behind other well-studied insects. In order to elucidate gene function across the order, it is necessary to establish tools that enable such gene manipulation. CRISPR/Cas9 is a promising technique and here we review the recent progress made in implementing the technique in butterflies; from broad patterning of the wing, to the development of specific colours in particular wing sections, to eyespot formation. The often species-specific responses to the CRISPR/Cas9-induced mutations in candidate genes, underscore the significance of these genes in the wide evolutionary

Advances in Insect Physiology, Volume 54
ISSN 0065-2806
https://doi.org/10.1016/bs.aiip.2017.11.001

diversification of butterfly wing patterns. We further discuss potential caveats in the interpretation of the resulting mutant phenotypes obtained through CRISPR/Cas9 gene editing. Finally, we discuss the possibilities CRISPR/Cas9 offers beyond mere knockout of candidate genes, including the potential for the generation of transgenics that will further elucidate the developmental genetic basis for wing pattern evolution.

ABBREVIATIONS

CDS coding DNA sequence
CRISPR clustered regularly interspaced short palindromic repeats
DSB double-stranded break
Indels insertion/deletions
KO(s) knockout(s)
RNAi RNA interference
sgRNA synthetic guide RNA
TALENs transcription activator-like effector nucleases
ZFNs zinc-finger nucleases

1. FUNCTIONAL GENOMICS IN LEPIDOPTERA: AN INTRODUCTION

The developmental evolution of butterfly wing patterns is a major model for the evolution of morphological novelties and the cooption of particular developmental genes therein (Brakefield et al., 1996; Deshmukh et al., 2017; Fujiwara and Nishikawa, 2016; Jiggins et al., 2017; Monteiro, 2015; Nijhout, 1991, 1996; Ohno and Otaki, 2015). Despite having identified many of the key genes involved over the years, and elucidated their expression patterns and allelic variability, we thus far lacked the tools to adequately study their function. Although some progress had been made in moths, in particular for the silk moth *Bombyx mori* (Kolliopoulou and Swevers, 2014), Lepidoptera as a whole are somewhat unique as a major model system in the field of evolutionary developmental biology to have been lacking such functional tools. Lepidopterans appear to have a very poor response to RNA interference (RNAi) (Kolliopoulou and Swevers, 2014; Terenius et al., 2011), which some have argued is a result of high levels of expression of dsRNases present endogenously within the order (Shukla et al., 2016), and because they display poor intracellular transport of the injected RNA (Joga et al., 2016). The development of RNAi and its application in moths have largely been driven by a desire to develop efficient pest control and safe guarding commercially interesting species such as *B. mori* (Kolliopoulou and Swevers, 2014). Very few moth studies have deployed RNAi in an evolutionary developmental biology context, and those that have done so use it in the context of early embryogenesis (Nakao, 2012, 2015). It is only very recently that two RNAi-based techniques showed promise for studying butterfly wing patterning and its evolution: (1) RNAi driven by an inducible heat-shock promoter in transgenic *Bicyclus anynana* butterflies (Monteiro et al., 2013; Ozsu et al., 2017) and (2) a novel RNAi technique (in combination with localised electroporation) in *Papilio* butterflies, which has been deployed in the study of wing patterning mimicry (Fujiwara and Nishikawa, 2016; Komata et al., 2016; Nishikawa et al., 2015).

Several other techniques, most notably zinc finger nucleases (ZFNs) and transcription activator-like effector nucleases (TALENs), are available that enable the study of gene function in insects through knockout (KO)/ knockdown, and again these have recently been developed in moths largely with a view to pest control; for example, disruption mating behaviour in the moth *Ostrinia furnacalis* by targeting odorant receptor coreceptor (*Orco*) (Yang et al., 2016). TALEN-mediated genome editing has indeed been well established now for *B. mori* (Daimon et al., 2014; Nakade et al., 2014; Takasu et al., 2010, 2013, 2016). For butterflies, few examples exist for the use of ZFNs. They were successfully employed to study the significance of *cryptochrome 2* for migration in the Monarch butterfly *Danaus plexippus* (Merlin et al., 2013), but the technique has not been deployed for the study of wing patterning. Both ZFNs and TALENs require the engineering of site-specific nucleases, which is time consuming, costly, and technically challenging (Gaj et al., 2013). The use of transposon-based transgenics has been used successfully in *B. mori* (Tamura et al., 2000) and has been extended to transgenic lines of *B. anynana* for evo-devo research via the PiggyBac-mediated integration (Marcus et al., 2004; Monteiro et al., 2013; Ozsu et al., 2017; Ramos and Monteiro, 2007; Tong et al., 2014). This approach, however, requires aggressive injection, screening, and line maintenance efforts that prevent its routine use for testing gene function. Finally, virus-mediated transformation techniques have been used with some success for gene overexpression in butterfly wings (Beldade and Peralta, 2017; Dhungel et al., 2013, 2016; Lewis et al., 1999), but here again, the difficulty of the method in question has prevented its routine use in spite of having yielded the first successful butterfly wing genetic modification results almost 20 years ago (Lewis et al., 1999).

The clustered regularly interspaced short palindromic repeat/CRISPR-associated (CRISPR/Cas9) system has recently been developed to study gene function. CRISPR/Cas9 is an RNA-guided nuclease tool for the targeted introduction of double-stranded DNA cleavage (Harrison et al., 2014; Hsu et al., 2014; Jinek et al., 2014) (Fig. 1). To knockout expression of the target genes, a synthetic guide RNA (sgRNA) is introduced into the desired organism, in conjunction with Cas9 protein or plasmid/capped mRNA encoding the protein (for a practical guide for Lepidoptera, see Zhang and Reed, 2017). Cas9 protein delivery appears to be an order of magnitude more efficient than mRNA delivery (Kistler et al., 2015). A 17–20 nt region in the sgRNA complementary to a portion of the (coding DNA sequence) CDS of the target gene binds the DNA and directs the recruitment of Cas9, an endonuclease which induces double-stranded breaks (DSBs) at the desired site (Hsu et al., 2014). These DSBs result in insertion/deletions (indels) at the targeted loci as a result of ineffective repair by nonhomologous end joining. If a coding exon is targeted, most of the cleavage events yield a frameshift resulting in nonfunctional proteins in those cells carrying the mutations. The power of the CRISPR/Cas9 system as compared to other gene-editing methods lies in its programmability: a single sgRNA is sufficient to guide the binding of Cas9, and the $N_{17-20}NGG$ (PAM) motif required for this binding is readily available within most target sequences (Hsu et al., 2014). Furthermore, the availability of a rapidly increasing number of available lepidopteran genomes (Challis et al., 2016), CRISPR/Cas9 has the tremendous potential to blur the boundaries between model and nonmodel organisms when it comes to detailed studies of gene function and their evolution. Indeed, CRISPR/Cas9 has been shown to work across a very wide range of organisms and with a wide variety of applications in mind (Barrangou and Doudna, 2016; Hsu et al., 2014).

In Lepidoptera, soon after its inception CRISPR/Cas9 was developed for *B. mori* (Daimon et al., 2014; Liu et al., 2014; Ma et al., 2014; Nakade et al., 2014; Wei et al., 2014). Given the commercial importance of *B. mori*, a novel combination of techniques has recently been developed to target a baculovirus in this species. It uses the PiggyBac transposon to integrate a transgene containing the Cas9 protein under the control of a constitutive *ie1* promoter (in the baculovirus), while the gRNA expression cassettes are under the control of U6 promoters. This construct prevents the baculovirus from disrupting its host's development (Chen et al., 2017). In general, CRISPR/Cas9 in moths is again mostly used in a pest control context. Just like for *O. furnacalis* (Yang et al., 2016), the odorant receptor coreceptor (*Orco*) has also been targeted in *Spodoptera littoralis* to disrupt sex pheromone and plant odour detection, in this case using CRISPR/Cas9 (Koutroumpa et al., 2016) rather than TALENs as was used in *O. furnacalis*. Disrupting pheromone communication as a means for pest control has been undertaken in *Helicoverpa armigera*, by targeting pheromone-binding protein

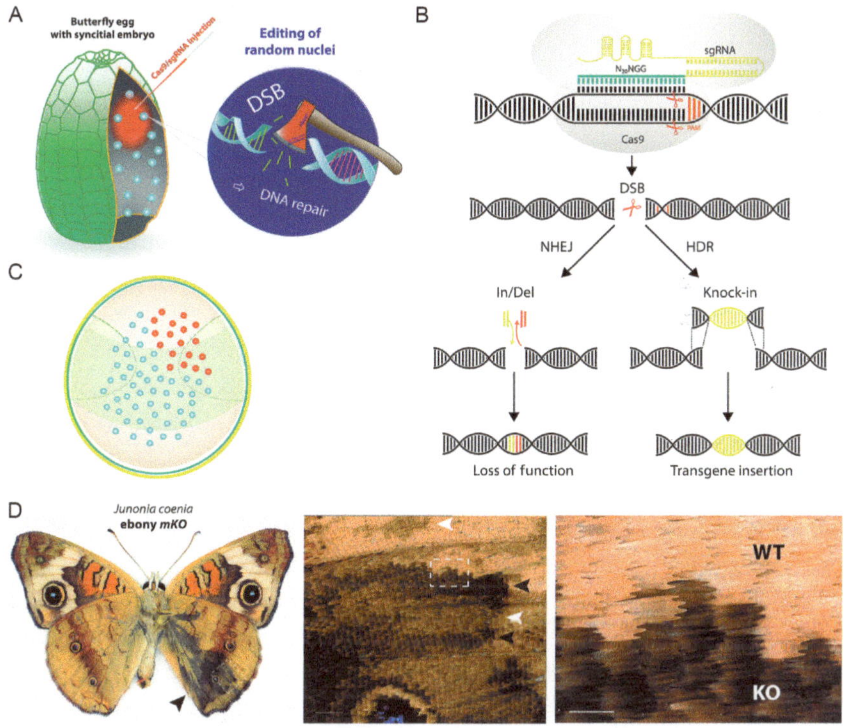

Fig. 1 Targeted CRISPR/Cas9 mutagenesis for the generation of mosaic gene knockouts (mKO) in butterflies. Cas9/sgRNA protein duplexes are injected into early stage syncytial embryos (which is in the first few hours after egg laying for butterflies), presumably entering a fraction of the dividing nuclei (A). RNA-guided cleavage induces a double-stranded DNA break three nucleotides upstream of the PAM sequence (*orange*), leading to DNA-repair pathways (B). The egg of a butterfly is maternally prepatterned into a region that is fate mapped to become the extraembryonic tissue called the serosa, and a region in the shape of a band (here in faint green) where the germ band and thus the embryo will form (Carter et al., 2015; Ferguson et al., 2014). Upon cellularisation after the syncytial stage, those nuclei carrying a CRISPR-induced mutation (*red*; wild type in *blue*) in this band region will result in an observable mosaic mutant phenotype in the developing butterfly (e.g. in the wings as discussed in this chapter) (C). Generation of hypermelanic wing clones in an adult buckeye butterfly, 1 month after the syncytial delivery CRISPR targeting the eumelanin inhibitor gene *ebony* (*e*) (D). *Arrowheads* show mutant clones of various identity, possibly reflecting two allelic states (*black*: $e^{-/-}$; *white*: $e^{+/-}$). Scale bar: 100 µm. *Panel (D): From Zhang, L.L., Martin, A., Perry, M.W., van der Burg, K.R.L., Matsuoka, Y., Monteiro, A., Reed, R.D., 2017. Genetic basis of melanin pigmentation in butterfly wings. Genetics 205, 1537–1550.*

(PBP1) (Ye et al., 2017) and the male pheromone antagonist Z11–16; OH (produced by females to avoid inopportune mating) (Chang et al., 2017) by means of CRISPR/Cas9. However, recently this species has been used to study pigmentation in Lepidoptera, by creating stable somatic and germline KO mutations of the ABC proteins, which transport pigment precursors from the cytoplasm (Khan et al., 2017). This study highlighted in particular the effects on eye pigmentation. Indeed, CRISPR/cas9 has also been used to investigate butterfly colour vision, by targeting rhodopsins in the Japanese yellow swallowtail (*Papilio xuthus*) and the painted lady (*Vanessa cardui*, Nymphalidae) (Perry et al., 2016). Given the fact that CRISPR/Cas9 is a relatively novel technique, the number of genes investigated in Lepidoptera is still small. In the context of embryonic and larval development the following genes have been targeted: the Hox genes *abdominal-A* (*abdA*) (*Plutella xylostella*, Huang et al., 2016; *Spodoptera litura*, Bi et al., 2016) and *Abdominal-B* (*AbdB*) (*P. xuthus* and *machaon*, Li et al., 2015, 2017) have been targeted, *wingless* (*B. mori*, Zhang et al., 2015), *frizzled* (*P. xuthus* and *machaon*, Li et al., 2015), *abnormal wing disc* and *fringe* (*B. mori*, Ling et al., 2015), and *FOXO* (*B. mori*, Zeng et al., 2017).

As can be expected for any technique that attempts to generate complete nulls of developmentally important genes (including ZFNs and TALENs), it is for the study of genes of importance in early embryogenesis, and larval growth, that this technique is not entirely without its limitations (at present). Introducing mutations early in development for key developmental genes is likely to be lethal to the animal before it reaches later developmental stages, as was shown, for example for *abd-A* and *Abd-B* (Bi et al., 2016; Huang et al., 2016; Li et al., 2015, 2017). Furthermore, interpreting mosaic phenotypes resulting from these injections is often challenging, as they result in partial phenotypes in the tissue of interest.

Although the technique of CRISPR/Cas9 may have some limitations and poses several technical challenges, it is currently the most promising technique available to facilitate the study of the functionality of candidate wing patterning genes in butterflies and relevant guidelines and protocols to that effect have been developed (Wang et al., 2015; Zhang and Reed, 2017). The developmental genes and their interactions underpinning wing patterning are at present still the most well-characterised aspect of development in butterflies (Deshmukh et al., 2017; Jiggins et al., 2017; Kunte et al., 2014; Monteiro, 2015; Monteiro et al., 2015; Nishikawa et al., 2015; Reed et al., 2011; Saenko et al., 2011; Wallbank et al., 2016). We will review here the insights we have gained, through CRISPR/Cas9, in the developmental evolution of the following genes involved in butterfly wing patterning: (1) the major prepatterning genes *WntA* (Mazo-Vargas et al., 2017) and *optix* (Zhang et al., 2017b), (2) pigmentation genes; *ebony* (Beldade and Peralta, 2017; Li et al., 2015), *yellow* (Perry et al., 2016; Zhang et al., 2017a) (as well as unpublished data on *Pararge aegeria*), *yellow-d*, *yellow-h2*, *yellow-h3*, *black*, *pale* (Zhang et al., 2017a), *Dopa decarboxylase Dcd* (Beldade and Peralta, 2017; Zhang et al., 2017a), and (3) genes involved in and affecting eyespot formation; *apterousA* (*apA*)/*apterousB* (*apB*) (Prakash and Monteiro, 2017), *spalt* (Zhang and Reed, 2016), and *Distal-less (Dll)* (Connahs et al., 2017; Zhang and Reed, 2016). We will then discuss where we should be aiming for in the field of the evolution and development of butterfly wing patterning. Ideally we should not only be aiming for knocking candidate genes out, but we should investigate specific polymorphisms of relevant genes in more detail, as well as the genes with which these genes interact and the control of their expression (e.g. through manipulation of regulatory regions). We will outline relevant approaches currently being developed.

2. INSIGHTS INTO BUTTERFLY WING PATTERNING AND EYESPOT DEVELOPMENT THROUGH CRISPR/CAS9

2.1 *WntA*: An Evolutionarily Malleable Genetic Pencil That Sketches Pattern Shapes

The gene *WntA* encodes a secreted ligand of the Wnt family, a molecular affiliation which makes it a good candidate for directing pattern formation processes. Indeed, its better-known relative *wingless* (*syn. Wnt1*) has been linked to pigment pattern formation in several insects (Koshikawa et al., 2015; Ozsu et al., 2017; Werner et al., 2010; Yamaguchi et al., 2013) and is a hallmark of developmental signalling. Nevertheless, it is not by a candidate gene approach that *WntA* was identified as a wing patterning factor, but by forward genetics: more precisely, by the crossing of multiple morphs in three species of *Heliconius* butterflies, each differing by the shape of their colour patterns, followed by the Mendelian mapping of this variation at the genetic level by chromosome walking (Martin et al., 2012; Papa et al., 2013). Later, the identification of additional *WntA* alleles was extended to more *Heliconius* populations and species by a combination of linkage and association mapping techniques (Huber et al., 2015; Van Belleghem et al., 2017), and more surprisingly, to the 65 My distant *Limenitis arthemis*, which displayed adaptive variation at the *WntA* locus across the US Northeast (Gallant et al., 2014). In total, no less than 18 alleles have been mapped as drivers of pattern shifts related to mimicry (Martin and Courtier-Orgogozo, 2017). Expression assays also suggested a key role for this gene in the induction of conserved pattern elements such as the Central Symmetry System, a stereotypical stripe that decorates the median region of many butterfly species (Martin et al., 2012; Mazo-Vargas et al., 2017). Nonetheless, it is only recently and with the use of CRISPR/Cas9 that this gene could be functionally validated. Mosaic KOs demonstrated an inductive role of *WntA* for the formation and positioning of many discrete pattern elements across seven species (Mazo-Vargas et al., 2017) (Table 1). We illustrate this principle here in the Gulf Fritillary butterfly, *Agraulis vanillae* (Fig. 2A). *WntA* shows an exquisitely complex expression pattern in last instar larval wing discs (Martin and Reed, 2014). In *A. vanillae*, it marks numerous spots corresponding to presumptive silver-spot patterns, each with an outline of black. Each *WntA*-positive silver pattern was lost or strongly reduced in the *WntA* KOs, demonstrating that *WntA* is necessary for the induction of these patterns (Mazo-Vargas et al., 2017). Seemingly *WntA*-negative silver-spots (outlined in yellow in Fig. 2A) expanded upon *WntA* KO, and conversely, contracted upon injection of heparin—a Wnt agonist (Martin and Reed, 2014; Mazo-Vargas et al., 2017). This suggests these upper hindwing silver spots are in fact inhibited by *WntA*, unlike the other silver spots of this butterfly. Thus depicting a complex picture of the *Agraulis* wing where identical patterns are under the antagonistic influence of the same pathway. While the mechanism behind this deserves further characterisation, it was observed that the *WntA*-negative patterns, in fact, are swept by a secondary wave of *WntA* expression in the late fifth instar stage, right before metamorphosis. From an evolutionary perspective, it is noteworthy that *Agraulis* patterns are thought to represent a highly derived, fragmented version of a more ancestral patterning system: the Nymphalid Ground Plan (NGP) (Nijhout and Wray, 1988). The combination of *WntA* expression and loss-of-function experiment validates this hypothesis. Indeed, in species representatives of the NGP, such as *Junonia coenia* (Fig. 2B), *V. cardui* (Fig. 2C), and *P. aegeria* (Fig. 2D), larval wing disc stripes of *WntA* expression delineates the position of signature components of the NGP such as the Central Symmetry System (in blue in Fig. 2B–D) and External Symmetry System (in green) (Table 1). Upon *WntA* KO, the Central Symmetry Systems were missing in the three species, and their External Symmetry Systems underwent a distal shift (Mazo-Vargas et al., 2017). Thus, *WntA* emerges as a key and conserved organising factor of the NGP, and underwent numerous changes in its expression to reshape the pattern boundaries and positions of more

Table 1 Summary of CRISPR Experiments Targeting Wing Patterning Genes to Date

Gene Targeted	Expression in Wing Tissue	Species Studied and Observed Mutant Phenotype
Dll[a,b]	5th instar: Eyespot centre and first ring + border of wing	Species studied: *Bicyclus anynana*, *Vanessa cardui*, and *Junonia coenia* *For B. anynana*: Exon 3 (Homeobox containing exon)—Missing eyespots, missing scales, lighter wing colouration, bisected eyespots, missing border patterning. Exon 2 (including 5′UTR)—Exon skipping producing a N-truncated protein with an intact homeobox, resulting in ectopic eyespots. For *V. cardui*: Increase in eyespot size or number. Effects on multiple noneyespot traits: parafocal and margin colour pattern elements, and pigmentation. Hyperpigmentation mutants and elongated eyespots observed in *J. coenia*
WntA[c]	5th instar: In Nymphalinae and Satyrinae: Basal, Central and Marginal symmetry systems. In Heliconiinae: Expression associated with black melanic regions in forewings and hindwings. In Danainae: Expression associated with wing vein margins	Species studied: *J. coenia*, *Pararge aegeria*, *V. cardui*, *Heliconius erato demophoon*, *H. sara sara*, *Agraulis vanillae*, and *Danaus plexippus* In Nymphalinae and Satyrinae: Loss of patterning in regions associated with *WntA* expression (i.e. Basal, Central and Marginal symmetry systems) Distalisation of distal parafocal elements. In *V. cardui*, loss of forewing eyespots In Heliconiinae: Distalisation of pigmented wing bands as a result of the loss of proximal pattern boundary in *WntA*-positive zone In the derived *A. vanillae*, both pattern inhibition and pattern expansion observed In Danainae: Expansion of white coloured scales in cells adjacent to wing veins
optix[d]	Early pupa	Species studied: *H. erato*, *A. vanillae*, *V. cardui*, and *J. coenia* Replacement of colour pigments (ommochromes) with melanin. Structural colouration defects, resulting in induction of blue iridescence in *J. coenia*. *Optix* appears to play a fundamental and deeply conserved role in regulating both pigmentary and structural coloration in butterfly wings
apterous (A and B)[e]	5th instar: Expressed uniformly in dorsal side of wing disc	Species studied: *B. anynana*. Severe mutants display complete lack of wings. Mosaic mutants display ventralisation of dorsal structures and patterns (note: Apterous A absent from future eyespot)

Continued

Table 1 Summary of CRISPR Experiments Targeting Wing Patterning Genes to Date—cont'd

Gene Targeted	Expression in Wing Tissue	Species Studied and Observed Mutant Phenotype
spalt[a]	Pupa	Species studied: *J. coenia* and *V. cardui*. Missing eyespots in both forewings and hindwings
ebony[f–h]	Late pupa	Species studied: *Papilio xuthus*, *B. anynana*, and *J. coenia*. Darkening of wing scales (depending on background of scales)
yellow[h,i]	Late pupa	Species studied: *P. aegeria*, *B. anynana*, and *V. cardui*. Lack of black melanisation in wing scales, which may extent to eyespots (*P. aegeria*)
Yellow-d[h]	Late pupa	Species studied: *V. cardui*. Promotion of orange–brown pigmentation in specific ventral forewing patterns
Ddc[a,g,h]	Late pupa	Species studied: *B. anynana* and *V. cardui*. Strong depigmentation of black scales
pale[h]	Late pupa	Species studied: *P. xuthus* and *J. coenia*. Cuticular defects, scale development defects, lack of melanisation in wing scales and thorax
black[h]	Late pupa	Species studied: *V. cardui*. Increased melanisation of wing scales and pupae

[a]Zhang and Reed (2016).
[b]Connahs et al. (2017).
[c]Mazo-Vargas et al. (2017).
[d]Zhang et al. (2017b).
[e]Prakash and Monteiro (2017).
[f]Li et al. (2015).
[g]Beldade and Peralta (2017).
[h]Perry et al. (2016).
[i]Zhang et al. (2017a).
Genes targeted along with respective expression in wing tissue and resulting phenotype are summarised.

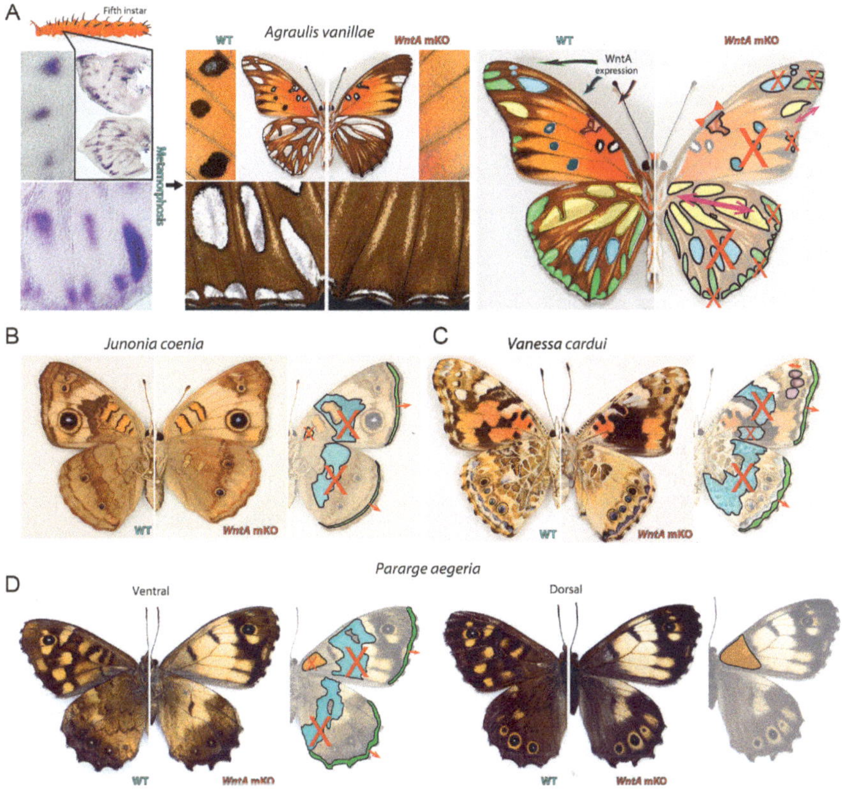

Fig. 2 Context dependent (species, wing region) inductive roles of *WntA* in nymphalid butterflies. Summary of *WntA* expression and KO effects on the ventral wing patterns of *Agraulis vanillae* (A), *Junonia coenia* (B), and *Vanessa cardui* (C). Panel D shows KO effects for both a ventral and dorsal of *P. aegeria*, highlighting the dorsal basal melanic patch on the right. See main text for details. Coloured layouts represent areas of larval *WntA* expression. *Symbols* show the effects of *WntA* loss-of-function (*crosses*: pattern removal; *red arrows*: contractions and shifts; *purple arrow*: expansion).

derived species such as *A. vanillae*, which underwent a drastic dislocation of its symmetry systems. This principle applies to even more derived nymphalid butterflies such as *Heliconius* and *Danaus* (Mazo-Vargas et al., 2017), which are not presented here for brevity.

2.2 CRISPR Mutants Highlight a Modular Organisation of Wing Patterns

A particularly salient outcome of the *WntA* loss-of-function phenotypes is the fact that they affect a fraction of discrete pattern elements. In other words, some pattern elements are *WntA* dependent, while other ones are

WntA independent, thus making apparent a highly modular architecture behind wing organisation. Allowing an uncoupling in the control of parts, modularity enables evolutionary flexibility, including plastic responses (Breuker et al., 2006; West-Eberhard, 2003). Here we illustrate this principle using serendipitous insights from the *WntA* mKO phenotypes of the Speckled Wood butterfly *P. aegeria*. Of interest here is a wing section which remains largely melanised while other areas become less melanised after KO; the dorsal basal forewing area (Fig. 2D). This in contrast to the corresponding ventral area. The dorso-basal patch thus appears *WntA* independent, an interesting finding from an evolutionary ecology perspective. Many butterflies, including *P. aegeria* (Berwaerts et al., 2001), rely on overall dorsal wing melanisation, but this basal part of the wing in particular, to heat up in order to become active (Wasserthal, 1975). *Pararge* wing pattern variation has been well documented (Schwanwitsch, 1935), and not only does there appear to be genetic variation between populations associated with melanisation (i.e. populations from Northern Europe being darker) (Vandewoestijne and Van Dyck, 2011), but there is also a large degree of developmental plasticity in melanisation in response to temperature and diet (Nylin et al., 1995), not least because the production of melanin is costly and subject to various resource trade-offs in this species (Talloen et al., 2004). One study concentrated on the size of the dorsal yellow areas in the border and central symmetry system, rather than the overall degree of melanisation. The size of the dorsal yellow areas has been associated with a degree of crypsis and thermoregulation, and associated differences in flight behaviours (Van Dyck and Matthysen, 1998). Males with less melanised wings and larger yellow patches spend significantly more time basking in sunlit patches, and darker males fly more frequently, suggesting an adaptive advantage to melanisation with respect to thermoregulation and patrolling behaviour (Van Dyck et al., 1997).

2.3 *optix*: A Master Regulator of Pigmentation and Iridescence

While *WntA* appears to instruct the formation of patterns associated with specific symmetry systems, another gene, *optix*, has also repeatedly been mapped in association with red and orange ommochrome colour patterns (Baxter et al., 2008; Huber et al., 2015; Reed et al., 2011; Van Belleghem et al., 2017). *WntA* encodes a signalling molecule, while *optix* encodes a homeobox transcription factor, which is expressed later in wing development, in a pattern that prefigures the red bands and stripes observed in many *Heliconius* species (Martin et al., 2014; Reed et al., 2011). Transcriptomic profiling by RNAseq in *V. cardui* showed an upregulation of *optix* in orange sections of the wings at the onset of pigment synthesis (Zhang et al., 2017a). Furthermore, *optix* is also expressed in wing areas associated with nonpigmentation traits in several species, such as derived scales hypothesised to be involved in wing coupling or in the spreading of male pheromones (Martin et al., 2014). Similar to *WntA* however, no functional evidence was available to demonstrate its direct involvement in ommochrome pigmentation or other aspects of scale cell determination until now.

Recently, CRISPR/Cas9 was used to knock *optix* out in four different species of butterflies to gauge its role in the development of wing patterning and its role in the evolutionary diversification of wing patterns. The species tested were *Heliconius erato*, *A. vanillae*, *V. cardui*, and *J. coenia* (Zhang et al., 2017b) (Table 1). The results confirmed that *optix* plays a fundamental role in ommochrome pattern development, where mosaic mutants displayed a complete replacement of ommochromes with melanins, resulting in largely black and grey butterflies (Table 1). Surprisingly, *optix* KOs revealed this function is probably not only shared among the *Heliconius* butterflies, but also among the more distantly related *V. cardui* and *J. coenia*, where the same phenotype was observed despite a divergence of ~75–80 Mya, suggesting a deep evolutionary origin for this trait in nymphalid butterflies.

Furthermore, the *optix* CRISPR mosaics had several structural abnormalities associated with specific scale types; the so-called wing conjugation scales. It appeared that *optix* KOs resulted in a complete transformation of wing conjugation scales to normal scales in all four species, illustrating that *optix* not only functions as an 'or' switch for ommochrome pigmentation, but also has a conserved role in scale type determination. Furthermore, in *J. coenia*, *optix* KOs induced strong colour iridescence in wing scales which are occupied by ommochrome pigments in the wild type, indicating that *optix* functions as a repressor of structural iridescence in this species. In summary, *optix* KOs reveal a conserved role for the gene in aspects of scale morphology and both structural and ommochrome colouration, which appears to have been repurposed throughout evolution to contribute to the diversity of wing patterns observed in nymphalid butterflies (Table 1).

The gene *optix* is not alone in having a hugely significant effect on the overall levels of melanin in the butterfly wing. An example is provided by that of the iconic case of industrial melanism in the peppered moth *Biston betularia* (Brakefield, 1990; Cook, 2003; Majerus, 1998). The overall degree of wing melanisation appears to be the in large part due to the *carbonaria* transposable element increasing the abundance of *cortex* transcript (van't Hof et al., 2016). This gene is a cell-cycle regulator during early wing disc development and which also has been implicated in the mimetic radiation in *Heliconius* butterflies (in particular via its effects on the yellow wing areas) (Nadeau et al., 2016). Presumably *cortex* plays a key role in the development of specific pigment containing scale types per se (Nadeau et al., 2016). The functionality of the gene *cortex* is thus still poorly understood and it is hoped that CRISPR/Cas9 can elucidate how a cell-cycle regulator in (early) wing development can have such a profound effect on wing patterning across both moths and butterflies.

2.4 Pigmentation Genes

The many cases of evolution of melanisation patterns per se, both at the micro- and macroevolutionary level, as well as developmental plasticity in melanin production (for example the aforementioned case of *P. aegeria*), have been textbook examples of evolution in action (Majerus, 1998). Although the key regulatory mechanisms may have been eluding us for a long time (van't Hof et al., 2011, 2016), the downstream genes involved in pigment production itself, in particular the melanin pathway, have become well characterised (True, 2003; Wright, 1987). To date, quite a few genes underlying the production of pigments in the wings, in particular in relation to melanisation, have been functionally tested using CRISPR/Cas9 in several butterfly species (Table 1): *J. coenia* (*pale* and *ebony*, Zhang et al., 2017a), *V. cardui* (*yellow*, *ebony*, *Ddc*, *black*, *yellow-d*, *yellow-h2*, and *yellow-h3*, Zhang et al., 2017a), *B. anynana* (*yellow*, *Ddc*, and *ebony*, Beldade and Peralta, 2017; Zhang et al., 2017a), *P. xuthus* (*pale* and *ebony*, Li et al., 2015; Zhang et al., 2017a), and we present here the results of *yellow* mutants in *P. aegeria*. *Pale*, *yellow*, and *Ddc* have been shown as melanin-promoting factors (Beldade and Peralta, 2017; Zhang et al., 2017a), *ebony*, *black*, and *yellow* affect the hue of colour patterns that do not appear to be typical black DOPA-derived eumelanins (Zhang et al., 2017a). Below, we will discuss the main findings for each of these genes in turn. A further two genes, *yellow-h2* and *yellow-h3* have also been functionally tested using CRISPR/Cas9, but mortality in the pupal stage prevented characterisation of adult pigment phenotypes (Zhang et al., 2017a).

The gene *pale* encodes the tyrosine hydroxylase enzyme, which catalyses the formation of dihydroxyphenylalanine (DOPA) from tyrosine at an early step of the melanin synthesis cascade (Wright, 1987). In *J. coenia*, *pale* mutants displayed severe cuticular defects including improperly expanded wings (Zhang et al., 2017a), presumably due to the role of DOPA in cuticle maturation and sclerotisation (Gorman and Arakane, 2010). Amelanism was also observed in *pale* mutants in thorax scales, wing scales, and in the inner ring scales of melanic eyespots, which became completely transparent (Zhang et al., 2017a) (Table 1).

The *yellow* gene family is large with a block of 3–5 genes highly conserved throughout insect diversification (Ferguson et al., 2011). Although the full function of many of the (pleiotropic) *yellow* genes is largely unknown (Ferguson et al., 2011), recent CRISPR/Cas9 functional analyses of *yellow*, *yellow-d*, *yellow-h2*, and *yellow-h3* have shown that these genes play a role in melanic pigmentation of butterfly wings (Perry et al., 2016; Zhang et al., 2017a). The gene *yellow* encodes a secreted extracellular protein required for production of black melanin pigments in *Drosophila* (Biessmann, 1985). KO of *yellow* in several species has been shown to result in loss of cuticular pigmentation (Wittkopp and Beldade, 2009). It has been shown in the butterfly *P. xuthus* (Futahashi and Fujiwara, 2007), for example, that 20-hydroxyecdysone (20E) hormone titres in the final larval instar regulate a number of (cuticular) pigmentation genes. A high titre of 20E leads to upregulation of *yellow*, whereas a low titre leads to downregulation of *ebony* and *Ddc*. Thus, cuticular pigmentation is significantly controlled by the exposure to 20E and its removal. The deployment of 20E in this context may very well be a mechanism underpinning the observed developmental plasticity in pigment production in many butterfly species as in-depth RNA-seq studies have indicated (Connahs et al., 2016; Daniels et al., 2014).

For *P. aegeria* mutations in *yellow* were produced by coinjecting two sgRNAs flanking the conserved major royal jelly protein (MRJP) motif. Examination of injected individuals revealed ectopic loss of black pigmentation across both forewings and hindwings (Fig. 3), in accordance with previously reported results in other butterfly species (Perry et al., 2016; Zhang et al., 2017a) (Table 1). The affected cells in the dorsal and ventral part of the wings were often asymmetrical, suggesting independent migration of affected cells during wing development. In the affected cells, areas of previous dark brown pigmentation became pale yellow, while black pigmented scales in the eyespots turned brown. Regions of yellow around the eyespots and in the forewings remained unaffected suggesting a crucial role for this gene in the synthesis of black melanin, while having no effect on lighter pigments of *P. aegeria*. Interestingly, darker areas on the ventral side of the forewings seemed to be differently affected in terms of colouration as compared

Fig. 3 Knockout of the *yellow* locus in *Pararge aegeria*. Wild-type *P. aegeria* (*left*) shown alongside a representative yellow KO individual (*right*) for both dorsal and ventral wing surfaces (A). The effects of the *yellow* KO were also observed in one of the hindwing eyespots (i.e. HW-ocellus 2, indicated by a *dashed box*) (B).

to the dorsal side, where the yellow around the hindwing eyespots is affected, while seeing no obvious effects on the yellow around the dorsal forewing eyespot. Loss of pigmentation in the abdomen and thorax of some individuals was also observed.

Through the microinjection of sgRNA complementary to the *yellow* CDS, mutations were introduced that resulted in cells carrying mKO phenotypes for the gene. In accordance with recently published results in other butterflies (Perry et al., 2016, Zhang et al., 2017a), *yellow* mKOs results in a loss of melanisation on the wings of *P. aegeria*. These results are also in accordance with expectations from phenotypes observed in *Drosophila* (Wittkopp et al., 2002), as well as other insects including *Tribolium castaneum* (Arakane et al., 2010), *Oncopeltus fasciatus* (Liu et al., 2016), and *B. mori* (Futahashi et al., 2008). The *yellow* KO results in *P. aegeria* thus support a conserved role for this gene being involved in melanin pigmentation. It is interesting to note that the mKO phenotypes were observed in both hindwings and forewings, as well as in the dorsal and ventral sides of the wings (Fig. 3). This is similar to the findings reported in *Drosophila*, but different to the RNAi phenotypes observed in *Tribolium* and *Oncopeltus*, where only hindwing defects were noted (Arakane et al., 2010; Liu et al., 2016). This discrepancy is also in accordance with a recently published study, where several species of butterflies with *yellow* mutants also contained both forewing and hindwing defects (Zhang et al., 2017a). These differences suggest that deployment of *yellow* has diverged throughout insect evolution. Furthermore, the effects of the *yellow* mKO did extend to the eyespots as well. Half of one eyespot on the hindwing (ocellus 2 in *P. aegeria*) lost its melanic scales in the area bordered by the unaffected yellow ring.

Zhang et al. (2017a) used CRISPR/Cas9 to functionally test the role of *yellow-d* in *V. cardui* wing scale pigmentation. Like *yellow*, *yellow-d* affected melanin patterns (Table 1). But unlike *yellow*, *yellow-d* also showed red-specific expression resulting in buff coloured pattern elements becoming a more chocolate-ochre hue. *Ddc* encodes the melanin synthesis enzyme DOPA decarboxylase (Wright, 1987). In *B. anynana*, CRISPR/Cas9-mediated *Ddc* KO in early embryos resulted in mosaic adults with pigment removed from the eyespot black disc (Beldade and Peralta, 2017) (Table 1). In *V. cardui*, *Ddc* mutants displayed pigment phenotypes similar to those observed for *yellow* (Zhang et al., 2017a). Unlike *yellow* however, *Ddc* KO in *V. cardui* also changed brown and tan pigments on the ventral wing surface (Zhang et al., 2017a). *V. cardui* larvae furthermore showed reduced melanin pigmentation, but *Ddc* KO typically resulted in high mortality, with many larvae failing to hatch from their eggs (Zhang et al., 2017a) (Table 1).

The gene *ebony* encodes the N-beta-alanyldopamine (NBAD) synthase enzyme, an important component of melanin biosynthesis (Wright, 1987), which is expressed in the larval, pupal, and adult stages of butterflies (Beldade and Peralta, 2017; Li et al., 2015; Zhang et al., 2017a). Mutant *ebony P. xuthus* larvae obtained with CRISPR/Cas9 showed enhanced melanic pigmentation (Li et al., 2015). In *V. cardui*, late-stage pupal wing phenotypes were observed whose red, white, and brown colour patterns produced an almost inverse colouration of the wild-type wing, but this phenotype could not be recovered in the adult stage due to high pupal mortality (Zhang et al., 2017a). KOs of *ebony* in *P. xuthus*, *B. anynana*, *J. coenia*, and *V. cardui* results in darkening of adult wing patches, evident across a range of colour pattern types including buffs, tans, and light and dark browns (Beldade and Peralta, 2017; Li et al., 2015; Zhang et al., 2017a) (Table 1).

Disruption of *black* leads to beta-alanine deficiencies, which limits NBAD scelerotin production resulting in an excess of dopamine, and causes cuticular melanisation (Wright, 1987). In *V. cardui* phenotypes with enhanced melanisation in pupal cuticles and wing scales was observed (Zhang et al., 2017a). In wing scales, *black* KO appeared to affect the same colour pattern elements as *yellow-d*, but without the marked yellow overtones (Zhang et al., 2017a) (Table 1 and Fig. 1D).

2.5 Genes Involved in Eyespot Development

Arguably, butterfly eyespots are one of the most well-studied wing pattern elements, with a large literature exploring their ecology, evolution, and development (Monteiro, 2015; Nijhout, 1991). To date, CRISPR/Cas9 has been used to study the function of three major transcription factor genes expressed in association with developing eyespots; *dll* (Connahs et al., 2017; Zhang and Reed, 2016), *spalt* (Zhang and Reed, 2016), and *apA/B* (Prakash and Monteiro, 2017) (Table 1). Previous functional analyses of the appendage-development gene *Dll* using germline transformation experiments in *B. anynana*, and baculovirus-mediated gene transfer experiments in *J. orithya* (Dhungel et al., 2016) demonstrated an eyespot-promoting role for this gene, a role which had been inferred before (Beldade et al., 2002; Carroll et al., 1994). The first use of CRISPR/Cas9 in an eyespot gene instead found an eyespot-repressing role for *Dll* (Zhang and Reed, 2016). Connahs et al. (2017) later solved that discrepancy by using CRISPR/Cas9 to target two different *Dll* exons. Disrupting the homeodomain in exon 3 produced missing eyespots, demonstrating its role as a positive eyespot regulator. In contrast, they found that targeting the exon 2 as what had been done in *V. cardui* produced a N-truncated Dll proteins with an intact homeodomain, which resulted in *B. anynana* butterflies with ectopic eyespots. This result is consistent with the previous study but changes its interpretation: the proper loss-of-function experiment is done by targeting exon 3 and reaffirms an eyespot activator role for *Dll*. It would appear that Zhang and Reed (2016) had actually generated a gain-of-function effect perhaps by truncating *Dll* from a repressor domain. Thus, these two studies offer an elegant case study showing that CRISPR somatic mutations can generate allelic heterogeneity depending on the target sites, a property that should induce caution, and also creativity among butterfly genome editors.

The transcription factor, *spalt* is one of the earliest factors expressed in eyespot foci (Brunetti et al., 2001). Zhang and Reed (2016) used CRSIPR/Cas9 to test the hypothesis that *spalt* is a regulator of eyespot colour patterns in *J. coenia* and *V. cardui* (Table 1). Deletion of *spalt* resulted in the loss and/or reduction of eyespots demonstrating that this gene plays a role in promoting eyespots (Zhang and Reed, 2016). Interestingly, a loss of an eyespot on one wing surface could occur without affecting an eyespot on the opposing surface, demonstrating that eyespot determination is dorsoventrally decoupled to a large degree (Zhang and Reed, 2016, and see *yellow* and *WntA* results for *P. aegeria* presented in Figs 2D and 3). The gene *apA*, the homologue of the dorso-ventral wing selector gene *ap* in *Drosophila*, is expressed on the dorsal surface of larval wings in *J. coenia* (Carroll et al., 1994). Expression of the paralogue *apB*, and the role of both homologues in wing development and patterning was unknown. Using CRISPR/Cas9, Prakash and Monteiro (2017) functionally tested the role of *ap* by disrupting the homeodomain and LIM domain of the homologue *apA*, and the LIM domain of *apB*. This study demonstrated that *apA* functions as both a repressor and modifier of ventral wing colour patterns, as well as a promoter of dorsal sexual ornaments in males (Table 1). In contrast however, no striking transformations of dorsal to ventral identity were observed in *apB* mutants, suggesting that only the *apA* homologue functions as a dorsal surface selector in *B. anynana*. Furthermore, *apA* has been implicated as a repressor of eyespot development in *B. anynana*; *apA* KO mutants had additional eyespots on the dorsal surface of hindwings (Prakash and Monteiro, 2017).

3. BENEFITS AND CAVEATS OF MOSAICISM

A demonstrably high biallelic KO rate makes it potentially feasible to characterise mutant phenotypes in injected individuals. Mosaicism is an essential property of the technique (Fig. 1C), as it allows some 'escaper' individuals to survive from the potentially deleterious effects of a given KO during embryogenesis and larval life, and yield healthy adults with a mutant (wing) phenotype. This inevitably comes with caveats: in short mosaicism prevents the *rigorous* (1) spatial delineation of mutant clones and (2) determination of their allelic states (and thus whether an allele is dominant or recessive). A pigment synthesis gene is expected to show *cell autonomous* effects, and in the case of *ebony* in *J. coenia* (Fig. 1D), it is even possible to see two degrees of hypermelanism that likely reflect a mosaic of heterozygous and homozygous mutant states. But making such assumptions is more problematic for developmental regulatory genes, which could show (1) *noncell autonomous* effects (propagation of phenotypic effects in genotypically wild-type cells), and idiosyncratic dominant, recessive, or other allele-dependent effects. Based on the rugged and elongated shape of iridescent *optix* mutant clones in *J. coenia* (comparable to the shape of *ebony* mutants) (Zhang et al., 2017b), we can reasonably assume this gene is *cell autonomous*. In contrast, *WntA* mutants show more rounded boundaries, or can affect patterns at a distance. For instance, parafocal elements were pushed towards the wing margin in *WntA* mutants while they do not show WntA expression (Fig. 4). These outcomes suggest a cell autonomous effect, as could be expected from a signalling molecule.

Connahs et al. (2017) took advantage of the visible cell autonomy of *Dll* KOs on scale pigmentation to map the spatial boundaries of the mutant clones on the wing, and implement the inferred initial conditions to test their mathematical model of eyespot pattern formation. Here again, the mosaicism was taken as an advantage. A particularly salient result that deserves attention is that when they attempted to genotype individual mutant and wing clones from a same surface by deep sequencing (MiSeq), they could not recover well-separated alleles. Based on this preliminary result and on our own experience, we now believe the direct genotyping of wing tissues cannot be used to map spatial clone boundaries or determine allelic state. There are several possible explanations for this. First, genotyping requires the amplification of DNA, which could capture heterogeneous cell populations irrelevant to the differentiation of wing scales, such as interstitial epithelial cells, neurons, or even circulating cells such as haemocytes. Second, it is possible that the

Fig. 4 Region-specific and distal effects of *WntA* mKO. Example of a *V. cardui WntA* mosaic mutant shown on its ventral side. The right half of this individual (left side of the picture) is wild type; its left forewing and hindwings are wild type anteriorly to M2, and mutant posteriorly to M2 (A). Arrowheads denote WT and mutant parafocal elements above (*white*) and below (*black*) the M2 vein. Displacement effect of *WntA* mKO on the *grey–blue* parafocal elements of *V. cardui* (*right*) compared to wild type (*left*), here in ventral hindwings (B). *WntA* is expressed at the margin in last instar larval wing discs and likely organises distal patterns via morphogenetic signalling activities.

epithelial cells of butterfly wings rapidly delaminate before being absorbed by the insect body, as described in *Drosophila* (Togel et al., 2008). Given these concerns, and at the same time the remarkable heuristic power of targeted mutagenesis, we advise both caution and creativity in the interpretation of phenotypic and genotypic data derived from mosaic mutants.

4. BEYOND KNOCKOUTS OF CANDIDATE GENES

CRISPR/Cas9-mediated KOs have been a major step forward for those wishing to elucidate the functional role of a number of candidate genes. However, KOs are only the beginning, and a number of research groups are developing CRISPR/Cas9-based techniques to go beyond KOs. For example, as discussed elsewhere in this chapter, mosaic phenotypes may not always be easily interpretable and/or informative of function and in such cases it is desirable to generate stable transgenic lines. A potential positive side effect of generating stable transgenic lines from CRISPR/CAS9-induced mosaic founders is the ability to recover an allelic series of the targeted locus. Mosaicism in germline of the injected founders can lead to transmission of an array of mutant alleles in the next generation (Wei et al., 2014; Yen et al., 2014). In butterflies, the successful transmission of an allelic series of mutants was first demonstrated in *D. plexippus* using ZFNs (Merlin et al., 2013). Germline transformation in general has also been achieved with *Pieris rapae* (Stoehr et al., 2015) and *B. anynana*, using a PiggyBac construct (Chen et al., 2011). The complexity of such germline mutations could be further increased by coinjecting a cocktail of several different sgRNAs targeting the same locus. The utility of an allelic series of protein-coding genes may not be intuitive; however, it could provide a crude but efficient tool to probe the function of noncoding regulatory loci. This approach has been used to functionally validate upstream regulatory elements of the *tyrosinase* gene in mice (Seruggia et al., 2015).

Such an approach might be feasible to begin dissecting the *cis*-regulatory elements associated with the reported red pigment patterns in *Heliconius* species (Nadeau et al., 2012, 2013; Reed et al., 2011; Wallbank et al., 2016). Genomic studies in *H. melpomene* and its relatives have further narrowed the location of these noncoding regulatory elements down to a few kilobases (Wallbank et al., 2016), and chromatin accessibility profiles and methods are now available in *Heliconius* to narrow down sites that are differentially active between tissues and stages (Lewis et al., 2016). Targeting these regions simultaneously with several sgRNAs, spaced out across the length of the locus, could potentially generate butterflies carrying regulatory alleles of various lengths (including complete deletions) (Fig. 5). Phenotypic analysis of this panel of alleles in heterozygous (F_1 generation) and/or in homozygous (F_2 generation) state could aid in identifying the functional elements of these regulatory regions and further refining the location of the causative regions underlying differences in red patterns. Such an approach could also be applied to putative regulatory regions of *WntA* (Martin et al., 2012) (Fig. 5). Moreover, targeting regulatory regions of the genes, instead of

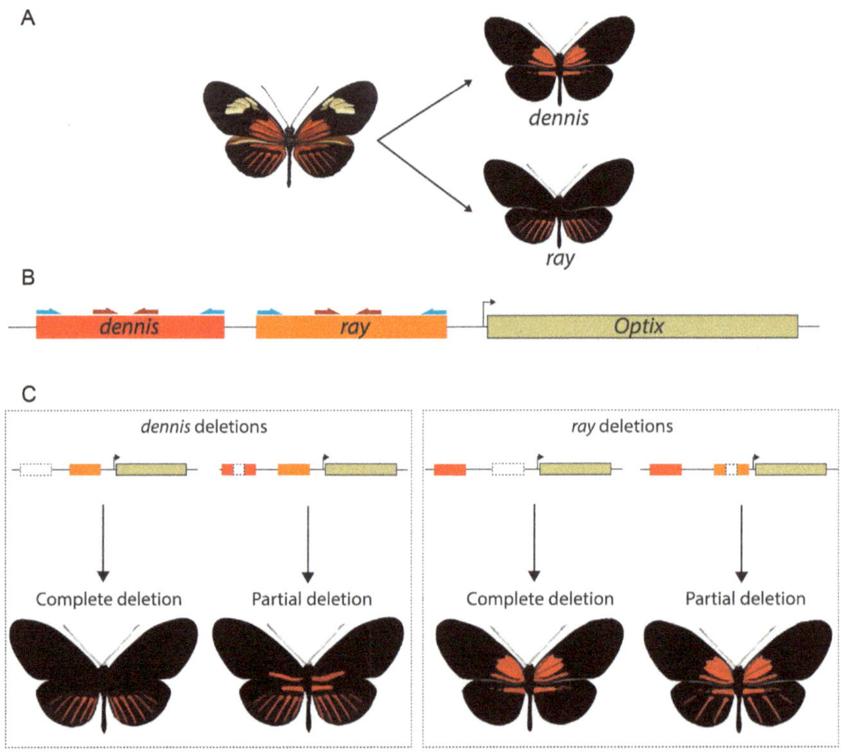

Fig. 5 The *optix* regulatory elements experiment. The red bands in many *Heliconius* species are decoupled into two developmental modules known as the *dennis* and *ray* phenotypes (A). These are controlled by two separate *cis*-regulatory regions located upstream of the *optix* gene (B). By inducing different deletions using combinations of sgRNAs (*red* and *blue arrows* in B), the contribution of each enhancer to colour pattern formation could in theory be established (C). Complete enhancer deletions should result the abolishment of each respective *dennis* or *ray* phenotype, while partial deletions would allow further dissection of the *cis*-regulatory logic controlling red pattern formation. Such a phenotype analysis would be effective if same/similar alleles could be brought together in homozygous state in an F_2 generation.

the protein-coding regions, potentially alleviates issues of lethality associated with complete functional KO of pleiotropic and early embryonic genes (Huang et al., 2016). However, it should be borne in mind that genetic variation and enhancer dominance may limit to an extent obtaining quick and easy-to-interpret results.

More precise manipulations such as integrating inducible expression cassettes; for conditional expression (Chen et al., 2011), insertion of small protein tags (e.g. V5 epitope), or larger fluorescent proteins for expression assays, and direct allelic replacement (to identify causative nucleotide(s)) would also benefit from the generation of stable transgenic lines. Such applications rely on using the CRISPR/CAS9 system alongside the cells' innate HDR machinery which has shown to be conducive for successfully generating knock-ins in butterflies (Zhang and Reed, 2017). These initial efficiencies of HDR-mediated knock-ins appear to be low but could potentially be increased by in vivo linearisation of donor DNA (Irion et al., 2014), the use of single-stranded oligodeoxynucleotides (ssODNs are particularly useful for introducing small changes (Ran et al., 2013), including the Easi-CRISPR technique (Quadros et al., 2017)), suppression of the innate NHEJ machinery (NHEJ and HDR machinery compete to repair DSBs; Singh et al., 2015) and perhaps even avoiding the HDR machinery altogether (Auer et al., 2014). It is likely that finetuning of these and other parameters are likely to lead to a further increase in knock-in efficiency in butterflies. In the future, the application of such experimental strategies may largely be limited by generation times, mating strategies, maintenance of inbred lines, and the ability to rear large numbers in the lab—factors that will vary by species.

The technique of CRISPR/Cas9 has proved a breakthrough for the study of butterfly wing pattern and eyespot development, and their evolution. Exciting new results are being produced at an accelerated rate and further advances of the technique, coupled with other techniques such as transcriptomics to identify novel relevant genes, will rapidly provide us with detailed insights into how the wide variety of beautiful butterfly wings develop and evolve.

ACKNOWLEDGEMENTS

Support for this research was provided by two Nigel Groome PhD studentships awarded to Breuker for Livraghi and Braak, and a NSF award to Martin (NSF-IOS #1656553). We thank Camille Tulure and Hans Van Dyck for providing *P. aegeria* animals to establish the breeding stocks that were used for *yellow* and *WntA* CRISPR/Cas9. We thank the George Washington University for assistance with figure design.

REFERENCES

Arakane, Y., Dittmer, N.T., Tomoyasu, Y., Kramer, K.J., Muthukrishnan, S., Beeman, R.W., Kanost, M.R., 2010. Identification, mRNA expression and functional analysis of several *yellow* family genes in *Tribolium castaneum*. Insect Biochem. Mol. Biol. 40, 259–266.

Auer, T.O., Duroure, K., De Cian, A., Concordet, J.P., Del Bene, F., 2014. Highly efficient CRISPR/Cas9-mediated knock-in in zebrafish by homology-independent DNA repair. Genome Res. 24, 142–153.

Barrangou, R., Doudna, J.A., 2016. Applications of CRISPR technologies in research and beyond. Nat. Biotechnol. 34, 933–941.

Baxter, S.W., Papa, R., Chamberlain, N., Humphray, S.J., Joron, M., Morrison, C., Ffrench-Constant, R.H., McMillan, W.O., Jiggins, C.D., 2008. Convergent evolution in the genetic basis of Mullerian mimicry in *Heliconius* butterflies. Genetics 180, 1567–1577.

Beldade, P., Peralta, C.M., 2017. Developmental and evolutionary mechanisms shaping butterfly eyespots. Curr. Opin. Insect Sci. 19, 22–29.

Beldade, P., Brakefield, P.M., Long, A.D., 2002. Contribution of *Distal-less* to quantitative variation in butterfly eyespots. Nature 415, 315–318.

Berwaerts, K., Van Dyck, H., Vints, E., Matthysen, E., 2001. Effect of manipulated wing characteristics and basking posture on thermal properties of the butterfly *Pararge aegeria* (L.). J. Zool. 255, 261–267.

Bi, H.L., Xu, J., Tan, A.J., Huang, Y.P., 2016. CRISPR/Cas9-mediated targeted gene mutagenesis in *Spodoptera litura*. Insect Sci. 23, 469–477.

Biessmann, H., 1985. Molecular analysis of the *yellow* gene (*Y*) region of *Drosophila melanogaster*. Proc. Natl. Acad. Sci. U.S.A. 82, 7369–7373.

Brakefield, P.M., 1990. A decline of melanism in the peppered moth *Biston betularia* in the Netherlands. Biol. J. Linn. Soc. 39, 327–334.

Brakefield, P.M., Gates, J., Keys, D., Kesbeke, F., Wijngaarden, P.J., Monteiro, A., French, V., Carroll, S.B., 1996. Development, plasticity and evolution of butterfly eyespot patterns. Nature 384, 236–242.

Breuker, C.J., Debat, V., Klingenberg, C.P., 2006. Functional evo-devo. Trends Ecol. Evol. 21, 488–492.

Brunetti, C.R., Selegue, J.E., Monteiro, A., French, V., Brakefield, P.M., Carroll, S.B., 2001. The generation and diversification of butterfly eyespot color patterns. Curr. Biol. 11, 1578–1585.

Carroll, S.B., Gates, J., Keys, D.N., Paddock, S.W., Panganiban, G.E.F., Selegue, J.E., Williams, J.A., 1994. Pattern formation and eyespot determination in butterfly wings. Science 265, 109–114.

Carter, J.M., Gibbs, M., Breuker, C.J., 2015. Divergent RNA localisation patterns of maternal genes regulating embryonic patterning in the butterfly *Pararge aegeria*. PLoS One 10, e0144471.

Challis, R.J., Kumar, S., Dasmahapatra, K.K.K., Jiggins, C.D., Blaxter, M., 2016. Lepbase: the lepidopteran genome database. bioRxiv, 056994. https://doi.org/10.1101/056994.

Chang, H.T., Liu, Y., Ai, D., Jiang, X.C., Dong, S.L., Wang, G.R., 2017. A pheromone antagonist regulates optimal mating time in the moth *Helicoverpa armigera*. Curr. Biol. 27, 1610–1615.

Chen, B., Hrycaj, S., Schinko, J.B., Podlaha, O., Wimmer, E.A., Popadic, A., Monteiro, A., 2011. *Pogostick*: a new versatile *piggyBac* vector for inducible gene over-expression and down-regulation in emerging model systems. PLoS One 6, e18659.

Chen, S., Hou, C., Bi, H., Wang, Y., Xu, J., Li, M., James, A.A., Huang, Y., Tan, A., 2017. Transgenic clustered regularly interspaced short palindromic repeat/Cas9-mediated viral gene targeting for antiviral therapy of *Bombyx mori* nucleopolyhedrovirus. J. Virol. 91, e02465-02416.

Connahs, H., Rhen, T., Simmons, R.B., 2016. Transcriptome analysis of the painted lady butterfly, *Vanessa cardui* during wing color pattern development. BMC Genomics 17, 270.

Connahs, H., Tlili, S., van Creij, J., Loo, T.Y.J., Banerjee, T., Saunders, T.E., Monteiro, A., 2017. Disrupting different *Distal-less* exons leads to ectopic and missing eyespots accurately modeled by reaction-diffusion mechanisms. bioRxiv, 183491. https://doi.org/10.1101/183491.

Cook, L.M., 2003. The rise and fall of the *Carbonaria* form of the peppered moth. Q. Rev. Biol. 78, 399–417.

Daimon, T., Kiuchi, T., Takasu, Y., 2014. Recent progress in genome engineering techniques in the silkworm, *Bombyx mori*. Dev. Growth Differ. 56, 14–25.

Daniels, E.V., Murad, R., Mortazavi, A., Reed, R.D., 2014. Extensive transcriptional response associated with seasonal plasticity of butterfly wing patterns. Mol. Ecol. 23, 6123–6134.

Deshmukh, R., Baral, S., Gandhimathi, A., Kuwalekar, M., Kunte, K., 2017. Mimicry in butterflies: co-option and a bag of magnificent developmental genetic tricks. Wiley Interdiscip. Rev. Dev. Biol, e291. https://doi.org/10.1002/wdev.291.

Dhungel, B., Ohno, Y., Matayoshi, R., Otaki, J.M., 2013. Baculovirus-mediated gene transfer in butterfly wings in vivo: an efficient expression system with an anti-gp64 antibody. BMC Biotechnol. 13, 9.

Dhungel, B., Ohno, Y., Matayoshi, R., Iwasaki, M., Taira, W., Adhikari, K., Gurung, R., Otaki, J.M., 2016. Distal-less induces elemental color patterns in *Junonia* butterfly wings. Zool. Lett. 2, 4.

Ferguson, L.C., Green, J., Surridge, A., Jiggins, C.D., 2011. Evolution of the insect *yellow* gene family. Mol. Biol. Evol. 28, 257–272.

Ferguson, L., Marletaz, F., Carter, J.-M., Taylor, W.R., Gibbs, M., Breuker, C.J., Holland, P.W.H., 2014. Ancient expansion of the Hox cluster in Lepidoptera generated four Homeobox genes implicated in extraembryonic tissue formation. PLoS Genet. 10, e1004698.

Fujiwara, H., Nishikawa, H., 2016. Functional analysis of genes involved in color pattern formation in Lepidoptera. Curr. Opin. Insect Sci. 17, 16–23.

Futahashi, R., Fujiwara, H., 2007. Regulation of 20-hydroxyecdysone on the larval pigmentation and the expression of melanin synthesis enzymes and *yellow* gene of the swallowtail butterfly, *Papilio xuthus*. Insect Biochem. Mol. Biol. 37, 855–864.

Futahashi, R., Sato, J., Meng, Y., Okamoto, S., Daimon, T., Yamamoto, K., Suetsugu, Y., Narukawa, J., Takahashi, H., Banno, Y., Katsuma, S., Shimada, T., Mita, K., Fujiwara, H., 2008. *yellow* and *ebony* are the responsible genes for the larval color mutants of the silkworm *Bombyx mori*. Genetics 180, 1995–2005.

Gaj, T., Gersbach, C.A., Barbas, C.F., 2013. ZFN, TALEN, and CRISPR/Cas-based methods for genome engineering. Trends Biotechnol. 31, 397–405.

Gallant, J.R., Imhoff, V.E., Martin, A., Savage, W.K., Chamberlain, N.L., Pote, B.L., Peterson, C., Smith, G.E., Evans, B., Reed, R.D., Kronforst, M.R., Mullen, S.P., 2014. Ancient homology underlies adaptive mimetic diversity across butterflies. Nat. Commun. 5, 4817.

Gorman, M.J., Arakane, Y., 2010. Tyrosine hydroxylase is required for cuticle sclerotization and pigmentation in *Tribolium castaneum*. Insect Biochem. Mol. Biol. 40, 267–273.

Harrison, M.M., Jenkins, B.V., O'Connor-Giles, K.M., Wildonger, J., 2014. A CRISPR view of development. Genes Dev. 28, 1859–1872.

Hsu, P.D., Lander, E.S., Zhang, F., 2014. Development and applications of CRISPR-Cas9 for genome engineering. Cell 157, 1262–1278.

Huang, Y.P., Chen, Y.Z., Zeng, B.S., Wang, Y.J., James, A.A., Gurr, G.M., Yang, G., Lin, X.J., You, M.S., 2016. CRISPR/Cas9 mediated knockout of the *abdominal-A* homeotic gene in the global pest, diamondback moth (*Plutella xylostella*). Insect Biochem. Mol. Biol. 75, 98–106.

Huber, B., Whibley, A., Poul, Y.L., Navarro, N., Martin, A., Baxter, S., Shah, A., Gilles, B., Wirth, T., McMillan, W.O., Joron, M., 2015. Conservatism and novelty in the genetic architecture of adaptation in *Heliconius* butterflies. Heredity 114, 515–524.

Irion, U., Krauss, J., Nusslein-Volhard, C., 2014. Precise and efficient genome editing in zebrafish using the CRISPR/Cas9 system. Development 141, 4827–4830.

Jiggins, C.D., Wallbank, R.W.R., Hanly, J.J., 2017. Waiting in the wings: what can we learn about gene co-option from the diversification of butterfly wing patterns? Philos. Trans. R. Soc. Lond. B Biol. Sci. 372, 20150485.

Jinek, M., Jiang, F.G., Taylor, D.W., Sternberg, S.H., Kaya, E., Ma, E.B., Anders, C., Hauer, M., Zhou, K.H., Lin, S., Kaplan, M., Iavarone, A.T., Charpentier, E., Nogales, E., Doudna, J.A., 2014. Structures of Cas9 endonucleases reveal RNA-mediated conformational activation. Science 343, 1247997.

Joga, M.R., Zotti, M.J., Smagghe, G., Christiaens, O., 2016. RNAi efficiency, systemic properties, and novel delivery methods for pest insect control: what we know so far. Front. Physiol. 7, 553.

Khan, S.A., Reichelt, M., Heckel, D.G., 2017. Functional analysis of the ABCs of eye color in *Helicoverpa armigera* with CRISPR/Cas9-induced mutations. Sci. Rep. 7, 40025.

Kistler, K.E., Vosshall, L.B., Matthews, B.J., 2015. Genome engineering with CRISPR-Cas9 in the mosquito *Aedes aegypti*. Cell Rep. 11, 51–60.

Kolliopoulou, A., Swevers, L., 2014. Recent progress in RNAi research in Lepidoptera: intracellular machinery, antiviral immune response and prospects for insect pest control. Curr. Opin. Insect Sci. 6, 28–34.

Komata, S., Lin, C.P., Iijima, T., Fujiwara, H., Sota, T., 2016. Identification of *doublesex* alleles associated with the female-limited Batesian mimicry polymorphism in *Papilio memnon*. Sci. Rep. 6, 34782.

Koshikawa, S., Giorgianni, M.W., Vaccaro, K., Kassner, V.A., Yoder, J.H., Werner, T., Carroll, S.B., 2015. Gain of *cis*-regulatory activities underlies novel domains of *wingless* gene expression in *Drosophila*. Proc. Natl. Acad. Sci. U.S.A. 112, 7524–7529.

Koutroumpa, F.A., Monsempes, C., Francois, M.C., de Cian, A., Royer, C., Concordet, J.P., Jacquin-Joly, E., 2016. Heritable genome editing with CRISPR/Cas9 induces anosmia in a crop pest moth. Sci. Rep. 6, 29620.

Kunte, K., Zhang, W., Tenger-Trolander, A., Palmer, D.H., Martin, A., Reed, R.D., Mullen, S.P., Kronforst, M.R., 2014. *doublesex* is a mimicry supergene. Nature 507, 229–232.

Lewis, D.L., DeCamillis, M.A., Brunetti, C.R., Halder, G., Kassner, V.A., Selegue, J.E., Higgs, S., Carroll, S.B., 1999. Ectopic gene expression and homeotic transformations in arthropods using recombinant *Sindbis* viruses. Curr. Biol. 9, 1279–1287.

Lewis, J.J., van der Burg, K.R.L., Mazo-Vargas, A., Reed, R.D., 2016. ChIP-Seq-annotated *Heliconius erato* genome highlights patterns of *cis*-regulatory evolution in Lepidoptera. Cell Rep. 16, 2855–2863.

Li, X., Fan, D., Zhang, W., Liu, G., Zhang, L., Zhao, L., Fang, X., Chen, L., Dong, Y., Chen, Y., Ding, Y., Zhao, R., Feng, M., Zhu, Y., Feng, Y., Jiang, X., Zhu, D., Xiang, H., Feng, X., Li, S., Wang, J., Zhang, G., Kronforst, M.R., Wang, W., 2015. Outbred genome sequencing and CRISPR/Cas9 gene editing in butterflies. Nat. Commun. 6, 8212.

Li, X.Y., Liu, G.C., Sheng, W.J., Dong, Z.W., Chen, L., Zhao, R.P., Wang, W., 2017. Genome editing in the butterfly type-species *Papilio machaon*. Insect Sci. 24, 708–711.

Ling, L., Ge, X., Li, Z., Zeng, B., Xu, J., Chen, X., Shang, P., James, A.A., Huang, Y., Tan, A., 2015. MiR-2 family targets awd and fng to regulate wing morphogenesis in *Bombyx mori*. RNA Biol. 12, 742–748.

Liu, Y., Ma, S., Wang, X., Chang, J., Gao, J., Shi, R., Zhang, J., Lu, W., Zhao, P., Xia, Q., 2014. Highly efficient multiplex targeted mutagenesis and genomic structure variation in *Bombyx mori* cells using CRISPR/Cas9. Insect Biochem. Mol. Biol. 49, 35–42.

Liu, J., Lemonds, T.R., Marden, J.H., Popadic, A., 2016. A pathway analysis of melanin patterning in a hemimetabolous insect. Genetics 203, 403–413.

Ma, S., Chang, J., Wang, X., Liu, Y., Zhang, J., Lu, W., Gao, J., Shi, R., Zhao, P., Xia, Q., 2014. CRISPR/Cas9 mediated multiplex genome editing and heritable mutagenesis of BmKu70 in *Bombyx mori*. Sci. Rep. 4, 4489.

Majerus, M.E.N., 1998. Melanism; Evolution in Action. Oxford University Press, New York.

Marcus, J.M., Ramos, D.M., Monteiro, A., 2004. Germline transformation of the butterfly *Bicyclus anynana*. Proc. R. Soc. Lond. Ser. B Biol. Sci. 271, S263–S265.

Martin, A., Courtier-Orgogozo, 2017. Morphological evolution repeatedly caused by mutations in signaling ligand genes. In: Sekimoto, T., Nijhout, F. (Eds.), Diversity and Evolution of Butterfly Wing Patterns: An Integrative Approach. Springer, Singapore.

Martin, A., Reed, R.D., 2014. Wnt signaling underlies evolution and development of the butterfly wing pattern symmetry systems. Dev. Biol. 395, 367–378.

Martin, A., Papa, R., Nadeau, N.J., Hill, R.I., Counterman, B.A., Halder, G., Jiggins, C.D., Kronforst, M.R., Long, A.D., McMillan, W.O., Reed, R.D., 2012. Diversification of complex butterfly wing patterns by repeated regulatory evolution of a Wnt ligand. Proc. Natl. Acad. Sci. U.S.A. 109, 12632–12637.

Martin, A., McCulloch, K.J., Patel, N.H., Briscoe, A.D., Gilbert, L.E., Reed, R.D., 2014. Multiple recent co-options of Optix associated with novel traits in adaptive butterfly wing radiations. EvoDevo 5, 7.

Mazo-Vargas, A., Concha, C., Livraghi, L., Massardo, D., Wallbank, R.W.R., Zhang, L., Papador, J.D., Martinez-Najera, D., Jiggins, C.D., Kronforst, M.R., Breuker, C.J., Reed, R.D., Patel, N.H., McMillan, W.O., Martin, A., 2017. Macroevolutionary shifts of WntA function potentiate butterfly wing-pattern diversity. Proc. Natl. Acad. Sci. U.S.A. 114, 10701–10706. https://doi.org/10.1073/pnas.1708149114.

Merlin, C., Beaver, L.E., Taylor, O.R., Wolfe, S.A., Reppert, S.M., 2013. Efficient targeted mutagenesis in the monarch butterfly using zinc-finger nucleases. Genome Res. 23, 159–168.

Monteiro, A., 2015. Origin, development, and evolution of butterfly eyespots. Annu. Rev. Entomol. 60, 253–271.

Monteiro, A., Chen, B., Ramos, D.M., Oliver, J.C., Tong, X.L., Guo, M., Wang, W.K., Fazzino, L., Kamal, F., 2013. Distal-less regulates eyespot patterns and melanization in *Bicyclus* butterflies. J. Exp. Zool. B Mol. Dev. Evol. 320B, 321–331.

Monteiro, A., Tong, X., Bear, A., Liew, S.F., Bhardwaj, S., Wasik, B.R., Dinwiddie, A., Bastianelli, C., Cheong, W.F., Wenk, M.R., Cao, H., Prudic, K.L., 2015. Differential expression of ecdysone receptor leads to variation in phenotypic plasticity across serial homologs. PLoS Genet. 11, e1005529.

Nadeau, N.J., Whibley, A., Jones, R.T., Davey, J.W., Dasmahapatra, K.K., Baxter, S.W., Quail, M.A., Joron, M., Ffrench-Constant, R.H., Blaxter, M.L., Mallet, J., Jiggins, C.D., 2012. Genomic islands of divergence in hybridizing *Heliconius* butterflies identified by large-scale targeted sequencing. Philos. Trans. R. Soc. Lond. B Biol. Sci. 367, 343–353.

Nadeau, N.J., Martin, S.H., Kozak, K.M., Salazar, C., Dasmahapatra, K.K., Davey, J.W., Baxter, S.W., Blaxter, M.L., Mallet, J., Jiggins, C.D., 2013. Genome-wide patterns of divergence and gene flow across a butterfly radiation. Mol. Ecol. 22, 814–826.

Nadeau, N.J., Pardo-Diaz, C., Whibley, A., Supple, M.A., Saenko, S.V., Wallbank, R.W.R., Wu, G.C., Maroja, L., Ferguson, L., Hanly, J.J., Hines, H., Salazar, C., Merrill, R.M., Dowling, A.J., Ffrench-Constant, R.H., Llaurens, V., Joron, M., McMillan, W.O., Jiggins, C.D., 2016. The gene *cortex* controls mimicry and crypsis in butterflies and moths. Nature 534, 106–110.

Nakade, S., Tsubota, T., Sakane, Y., Kume, S., Sakamoto, N., Obara, M., Daimon, T., Sezutsu, H., Yamamoto, T., Sakuma, T., Suzuki, K.T., 2014. Microhomology-mediated end-joining-dependent integration of donor DNA in cells and animals using TALENs and CRISPR/Cas9. Nat. Commun. 5, 5560.

Nakao, H., 2012. Anterior and posterior centers jointly regulate *Bombyx* embryo body segmentation. Dev. Biol. 371, 293–301.

Nakao, H., 2015. Analyses of interactions among pair-rule genes and the gap gene *Kruppel* in *Bombyx* segmentation. Dev. Biol. 405, 149–157.

Nijhout, H.F., 1991. The Development and Evolution of Butterfly Wing Patterns. Smithsonian Institution Press, Washington.

Nijhout, H.F., 1996. Focus on butterfly eyespot development. Nature 384, 209–210.

Nijhout, H.F., Wray, G.A., 1988. Homologies in the color patterns of the genus *Heliconius* (Lepidoptera, Nymphalidae). Biol. J. Linn. Soc. 33, 345–365.

Nishikawa, H., Iijima, T., Kajitani, R., Yamaguchi, J., Ando, T., Suzuki, Y., Sugano, S., Fujiyama, A., Kosugi, S., Hirakawa, H., Tabata, S., Ozaki, K., Morimoto, H., Ihara, K., Obara, M., Hori, H., Itoh, T., Fujiwara, H., 2015. A genetic mechanism for female-limited Batesian mimicry in *Papilio* butterfly. Nat. Genet. 47, 405–409.

Nylin, S., Wickman, P.-O., Wiklund, C., 1995. Life-cycle regulation and life history plasticity in the speckled wood butterfly: are reaction norms predictable? Biol. J. Linn. Soc. 55, 143–157.

Ohno, Y., Otaki, J.M., 2015. Spontaneous long-range calcium waves in developing butterfly wings. BMC Dev. Biol. 15, 17.

Ozsu, N., Chan, Q.Y., Chen, B., Gupta, M.D., Monteiro, A., 2017. Wingless is a positive regulator of eyespot color patterns in *Bicyclus anynana* butterflies. Dev. Biol. 429, 177–185.

Papa, R., Kapan, D.D., Counterman, B.A., Maldonado, K., Lindstrom, D.P., Reed, R.D., Nijhout, H.F., Hrbek, T., McMillan, W.O., 2013. Multi-allelic major effect genes interact with minor effect QTLs to control adaptive color pattern variation in *Heliconius erato*. PLoS One 8, e57033.

Perry, M., Kinoshita, M., Saldi, G., Huo, L., Arikawa, K., Desplan, C., 2016. Molecular logic behind the three-way stochastic choices that expand butterfly colour vision. Nature 535, 280–284.

Prakash, A., Monteiro, A., 2017. *apterous* specifies dorsal wing patterns and sexual traits in butterflies. bioRxiv, 131011. https://doi.org/10.1101/131011.

Quadros, R.M., Miura, H., Harms, D.W., Akatsuka, H., Sato, T., Aida, T., Redder, R., Richardson, G.P., Inagaki, Y., Sakai, D., Buckley, S.M., Seshacharyulu, P., Batra, S.K., Behlke, M.A., Zeiner, S.A., Jacobi, A.M., Izu, Y., Thoreson, W.B., Urness, L.D., Mansour, S.L., Ohtsuka, M., Gurumurthy, C.B., 2017. Easi-CRISPR: a robust method for one-step generation of mice carrying conditional and insertion alleles using long ssDNA donors and CRISPR ribonucleoproteins. Genome Biol. 18, 92.

Ramos, D.M., Monteiro, A., 2007. Transgenic approaches to study wing color pattern development in Lepidoptera. Mol. Biosyst. 3, 530–535.

Ran, F.A., Hsu, P.D., Wright, J., Agarwala, V., Scott, D.A., Zhang, F., 2013. Genome engineering using the CRISPR-Cas9 system. Nat. Protoc. 8, 2281–2308.

Reed, R.D., Papa, R., Martin, A., Hines, H.M., Counterman, B.A., Pardo-Diaz, C., Jiggins, C.D., Chamberlain, N.L., Kronforst, M.R., Chen, R., Halder, G., Nijhout, H.F., McMillan, W.O., 2011. *optix* drives the repeated convergent evolution of butterfly wing pattern mimicry. Science 333, 1137–1141.

Saenko, S., Marialva, M., Beldade, P., 2011. Involvement of the conserved *Hox* gene *Antennapedia* in the development and evolution of a novel trait. EvoDevo 2, 9.

Schwanwitsch, B.N., 1935. Evolution of the wing-pattern in palaearctic Satyridae. III. Genus *Pararge* and five others. Acta Zool. 16, 145–281.

Seruggia, D., Fernandez, A., Cantero, M., Pelczar, P., Montoliu, L., 2015. Functional validation of mouse tyrosinase non-coding regulatory DNA elements by CRISPR-Cas9-mediated mutagenesis. Nucleic Acids Res. 43, 4855–4867.

Shukla, J.N., Kalsi, M., Sethi, A., Narva, K.E., Fishilevich, E., Singh, S., Mogilicherla, K., Palli, S.R., 2016. Reduced stability and intracellular transport of dsRNA contribute to poor RNAi response in lepidopteran insects. RNA Biol. 13, 656–669.

Singh, P., Schimenti, J.C., Bolcun-Filas, E., 2015. A mouse geneticist's practical guide to CRISPR applications. Genetics 199, 1–15.

Stoehr, A.M., Tong, X., Podlahas, O., Monteiro, A., 2015. Progress towards germ-line transformation of the butterfly *Pieris rapae* L. (Lepidoptera: Pieridae). Gene Technol. 4, 123.

Takasu, Y., Kobayashi, I., Beumer, K., Uchino, K., Sezutsu, H., Sajwan, S., Carroll, D., Tamura, T., Zurovec, M., 2010. Targeted mutagenesis in the silkworm *Bombyx mori* using zinc finger nuclease mRNA injection. Insect Biochem. Mol. Biol. 40, 759–765.

Takasu, Y., Sajwan, S., Daimon, T., Osanai-Futahashi, M., Uchino, K., Sezutsu, H., Tamura, T., Zurovec, M., 2013. Efficient TALEN construction for *Bombyx mori* gene targeting. PLoS One 8, e73458.

Takasu, Y., Kobayashi, I., Tamura, T., Uchino, K., Sezutsu, H., Zurovec, M., 2016. Precise genome editing in the silkworm *Bombyx mori* using TALENs and ds- and ssDNA donors—a practical approach. Insect Biochem. Mol. Biol. 78, 29–38.

Talloen, W., van Dyck, H., Lens, L., 2004. The cost of melanization: butterfly wing coloration under environmental stress. Evolution 58, 360–366.

Tamura, T., Thilbert, C., Royer, C., Kanda, T., Abraham, E., Kamba, M., Komoto, N., Thomas, J.L., Mauchamp, B., Chavancy, G., Shirk, P., Fraser, M., Prudhomme, J.C., Couble, P., 2000. Germline transformation of the silkworm *Bombyx mori* L-using a piggyBac transposon-derived vector. Nat. Biotechnol. 18, 81–84.

Terenius, O., Papanicolaou, A., Garbutt, J.S., Eleftherianos, I., Huvenne, H., Kanginakudru, S., Albrechtsen, M., An, C.J., Aymeric, J.L., Barthel, A., Bebas, P., Bitra, K., Bravo, A., Chevalieri, F., Collinge, D.P., Crava, C.M., de Maagd, R.A., Duvic, B., Erlandson, M., Faye, I., Felfoldi, G., Fujiwara, H., Futahashi, R., Gandhe, A.S., Gatehouse, H.S., Gatehouse, L.N., Giebultowicz, J.M., Gomez, I., Grimmelikhuijzen, C.J.P., Groot, A.T., Hauser, F., Heckel, D.G., Hegedus, D.D., Hrycaj, S., Huang, L.H., Hull, J.J., Iatrou, K., Iga, M., Kanost, M.R., Kotwica, J., Li, C.Y., Li, J.H., Liu, J.S., Lundmark, M., Matsumoto, S., Meyering-Vos, M., Millichap, P.J., Monteiro, A., Mrinal, N., Niimi, T., Nowara, D., Ohnishi, A., Oostra, V., Ozaki, K., Papakonstantinou, M., Popadic, A., Rajam, M.V., Saenko, S., Simpson, R.M., Soberon, M., Strand, M.R., Tomita, S., Toprak, U., Wang, P., Wee, C.W., Whyard, S., Zhang, W.Q., Nagaraju, J., Ffrench-Constant, R.H., Herrero, S., Gordon, K., Swelters, L., Smagghe, G., 2011. RNA interference in Lepidoptera: an overview of successful and unsuccessful studies and implications for experimental design. J. Insect Physiol. 57, 231–245.

Togel, M., Pass, G., Paululat, A., 2008. The *Drosophila* wing hearts originate from pericardial cells and are essential for wing maturation. Dev. Biol. 318, 29–37.

Tong, X.L., Hrycaj, S., Podlaha, O., Popadic, A., Monteiro, A., 2014. Over-expression of *Ultrabithorax* alters embryonic body plan and wing patterns in the butterfly *Bicyclus anynana*. Dev. Biol. 394, 357–366.

True, J.R., 2003. Insect melanism: the molecules matter. Trends Ecol. Evol. 18, 640–647.

Van Belleghem, S.M., Rastas, P., Papanicolaou, A., Martin, S.H., Arias, C.F., Supple, M.A., Hanly, J.J., Mallet, J., Lewis, J.J., Hines, H.M., Ruiz, M., Salazar, C., Linares, M., Moreira, G.R.P., Jiggins, C.D., Counterman, B.A., McMillan, W.O., Papa, R., 2017. Complex modular architecture around a simple toolkit of wing pattern genes. Nat. Ecol. Evol. 1, 52.

Van Dyck, H., Matthysen, E., 1998. Thermoregulatory differences between phenotypes in the speckled wood butterfly: hot perchers and cold patrollers? Oecologia 114, 326–334.

Van Dyck, H., Matthysen, E., Dhondt, A.A., 1997. The effect of wing colour on male behavioural strategies in the speckled wood butterfly. Anim. Behav. 53, 39–51.

Vandewoestijne, S., Van Dyck, H., 2011. Flight morphology along a latitudinal gradient in a butterfly: do geographic clines differ between agricultural and woodland landscapes? Ecography 34, 876–886.

van't Hof, A.E., Edmonds, N., Dalikova, M., Marec, F., Saccheri, I.J., 2011. Industrial melanism in British peppered moths has a singular and recent mutational origin. Science 332, 958–960.

van't Hof, A.E., Campagne, P., Rigden, D.J., Yung, C.J., Lingley, J., Quail, M.A., Hall, N., Darby, A.C., Saccheri, I.J., 2016. The industrial melanism mutation in British peppered moths is a transposable element. Nature 534, 102–105.

Wallbank, R.W.R., Baxter, S.W., Pardo-Diaz, C., Hanly, J.J., Martin, S.H., Mallet, J., Dasmahapatra, K.K., Salazar, C., Joron, M., Nadeau, N., McMillan, W.O., Jiggins, C.D., 2016. Evolutionary novelty in a butterfly wing pattern through enhancer shuffling. PLoS Biol. 14, e1002353.

Wang, W., Li, X., Liu, G., Chen, L., 2015. CRISPR/Cas9 gene editing in the swallowtail butterfly *Papilio xuthus*. Protoc. Exch. https://doi.org/10.1038/protex.2015.089.

Wasserthal, L.T., 1975. Role of butterfly wings in regulation of body temperature. J. Insect Physiol. 21, 1921–1930.

Wei, W., Xin, H., Roy, B., Dai, J., Miao, Y., Gao, G., 2014. Heritable genome editing with CRISPR/Cas9 in the silkworm, Bombyx mori. PLoS One 9, e101210.

Werner, T., Koshikawa, S., Williams, T.M., Carroll, S.B., 2010. Generation of a novel wing colour pattern by the wingless morphogen. Nature 464, 1143–1148.

West-Eberhard, M.J., 2003. Modularity. In: Developmental Plasticity and Evolution. Oxford University Press, New York, pp. 56–88.

Wittkopp, P.J., Beldade, P., 2009. Development and evolution of insect pigmentation: genetic mechanisms and the potential consequences of pleiotropy. Semin. Cell Dev. Biol. 20, 65–71.

Wittkopp, P.J., Vaccaro, K., Carroll, S.B., 2002. Evolution of *yellow* gene regulation and pigmentation in *Drosophila*. Curr. Biol. 12, 1547–1556.

Wright, T.R.F., 1987. The genetics of biogenic-amine metabolism, sclerotisation, and melanization in *Drosophila melanogaster*. Adv. Genet. 24, 127–222.

Yamaguchi, J., Banno, Y., Mita, K., Yamamoto, K., Ando, T., Fujiwara, H., 2013. Periodic Wnt1 expression in response to ecdysteroid generates twin-spot markings on caterpillars. Nat. Commun. 4, 1857.

Yang, B., Fujii, T., Ishikawa, Y., Matsuo, T., 2016. Targeted mutagenesis of an odorant receptor co-receptor using TALEN in *Ostrinia furnacalis*. Insect Biochem. Mol. Biol. 70, 53–59.

Ye, Z.F., Liu, X.L., Han, Q., Liao, H., Dong, X.T., Zhu, G.H., Dong, S.L., 2017. Functional characterization of *PBP1* gene in *Helicoverpa armigera* (Lepidoptera: Noctuidae) by using the CRISPR/Cas9 system. Sci. Rep. 7, 8470.

Yen, S.T., Zhang, M., Deng, J.M., Usman, S.J., Smith, C.N., Parker-Thornburg, J., Swinton, P.G., Martin, J.F., Behringer, R.R., 2014. Somatic mosaicism and allele complexity induced by CRISPR/Cas9 RNA injections in mouse zygotes. Dev. Biol. 393, 3–9.

Zeng, B., Huang, Y., Xu, J., Shiotsuki, T., Bai, H., Palli, S.R., Tan, A., 2017. The FOXO transcription factor controls insect growth and development by regulating juvenile hormone degradation in the silkworm, *Bombyx mori*. J. Biol. Chem. 292, 11659–11669.

Zhang, L.L., Reed, R.D., 2016. Genome editing in butterflies reveals that *spalt* promotes and *Distal-less* represses eyespot colour patterns. Nat. Commun. 7, 11769.

Zhang, L., Reed, R., 2017. A practical guide to CRISPR/Cas9 genome editing in Lepidoptera. In: Sekimoto, T., Nijhout, F. (Eds.), Diversity and Evolution of Butterfly Wing Patterns: An Integrative Approach. Springer, Singapore.

Zhang, Z.J., Aslam, A.M., Liu, X.J., Li, M.W., Huang, Y.P., Tan, A.J., 2015. Functional analysis of *Bombyx* Wnt1 during embryogenesis using the CRISPR/Cas9 system. J. Insect Physiol. 79, 73–79.

Zhang, L.L., Martin, A., Perry, M.W., van der Burg, K.R.L., Matsuoka, Y., Monteiro, A., Reed, R.D., 2017a. Genetic basis of melanin pigmentation in butterfly wings. Genetics 205, 1537–1550.

Zhang, L., Mazo-Vargas, A., Reed, R.D., 2017b. Single master regulatory gene coordinates the evolution and development of butterfly color and iridescence. Proc. Natl. Acad. Sci. U.S.A. 114, 10707–10712. https://doi.org/10.1073/pnas.1709058114.

FURTHER READING

Sekimura, T., Nijhout, F., 2017. Diversity and Evolution of Butterfly Wing Patterns: An Integrated Approach. Springer, Singapore.